CIVIL AIRLINER RECOGNITION

Peter R. March

LONDON

IAN ALLAN LTD

Contents

Front cover:
BAe 146-200A operated by PSA. *Neil Hargreaves*

Back cover, top:
The Saab 34D is operated by Piedmont Commuter. *Saab*

Back cover, bottom:
The Boeing 737-300 can be distinguished by the engines projecting well forward of the wing. *Boeing*

Sole distributors for the USA

Osceola, Wisconsin 54020, USA ®

First published 1989

ISBN 0 7110 1820 0

Published by Ian Allan Ltd, Shepperton, Surrey; and printed by Ian Allan Printing Ltd at their works at Coombelands in Runnymede, England

All photographs Peter R. March (PRM) unless otherwise credited.

Introduction

This second, colour edition of *Civil Airliner Recognition* provides the airport visitor, the airline passenger and the interested bystander with a compact guide to the many and varied airliners likely to be seen over the skies of the western world. All of the major types are classified according to the number and type of engines as this is the first and foremost recognition feature. This is followed by basic technical information on one of the variants, the date that the first aircraft (usually the prototype) was flown, production details and an indication of some of the operators using this airliner either currently or in recent years. The main recognition features are listed and where there are several variants these are also noted. Photographs are used to aid recognition, and distinctive variants are also illustrated where possible.

Some older types of aircraft such as the Convair Coronado remain in service in very small numbers; while at the other end of the time-scale there are new variants and types in the early stages of development and not yet in full service. These old and new airliners have been covered briefly in the final section. To assist in locating specific aircraft by name and/or manufacturer an index and cross-reference has also been provided.

Acknowledgements
The author would like to thank Steven Byles, Wal Gandy and Edwin Shackleton for their help in checking the text and the following for the supply of photographs: Austin J. Brown, Paul Gingell, Neil Hargreaves, Andrew March, Daniel March, Colin Wood, Alan Wright, Roger Wright and the various airliner manufacturers.

Engines

Piston engines

The older piston-engined airliners, such as the 50-year-old Douglas DC-3, have radial engines. More modern aircraft are fitted with in-line piston engines such as the Avco Lycoming or Continental flat-six series.

Below left:
Pratt & Whitney R1830-92 radial piston engines on a Douglas DC-3

Below right:
A modern in-line piston engine — the Avco Lycoming O-540-E4C, here powering a Britten-Norman Trislander

Turbojets

Only a handful of aircraft fitted with turbojets remain in service today. Early marks of Caravelle have the Rolls-Royce Avon 532R turbojet.

Below:
Caravelle V1-N F-BVPZ, with Rolls-Royce Avon 532R turbojets

Turboprops

In the 1960s and 1970s the Rolls-Royce Dart turboprop was widely used for medium and short range airliners such as the Herald, Friendship and Gulfstream I. The early turboprops like the Dart and the Turbomeca Bastan have been replaced by a new generation of shorter, broader turboprops like the Pratt & Whitney PT6A and Garrett TPE 331 from the USA.

Below:
Rolls-Royce Dart RDa7-536 turboprops power the British Aerospace HS748

Below left:
Turbomeca Bastan VIC turboprop on a Nord 262

Below right:
Newer turboprops include the Pratt & Whitney PT6A-50, seen here on a DHC (Boeing) Dash 7

Turbofans

These have replaced the turbojets in most airliners – they are quieter, more powerful and generally have a larger diameter intake, shorter length and a small diameter jet pipe.

Below left:
General Electric/SNECMA CFM 56-2 turbofans power the McDonnell Douglas DC-8 series 70 aircraft

Below right:
Rolls-Royce RB211 turbofans on the Boeing 757

Below:
General Electric/SNECMA CFM56 turbofans on a Boeing 737-300

Four-engined jet airliners

Aerospatiale/BAC Concorde

Four turbojet supersonic airliner

Basic data for Concorde 102

Powerplant: Four Rolls-Royce/ SNECMA Olympus 593 (reheat) of 38,050lb st
Span: 83ft 10in (25.47m)
Length: 203ft 9in (62.10m)
Max cruise: 1,336mph (2,150km/hr)
Passengers: 128 plus three crew

First aircraft flown: 2 March 1969
Production: Two prototypes, two pre-production and 16 production aircraft

Recent/current service with: Air France (seven) and British Airways (seven)
Recognition: Underwing mounted engines in two double nacelles. Slender, low-set delta wing. Narrow fuselage with pointed nose that droops for landing. No tailplane, angular fin and rudder. Very tall stalky undercarriage
Variants: None amongst the aircraft in service

Below:
An Aerospatiale/BAC Concorde in British Airways service.

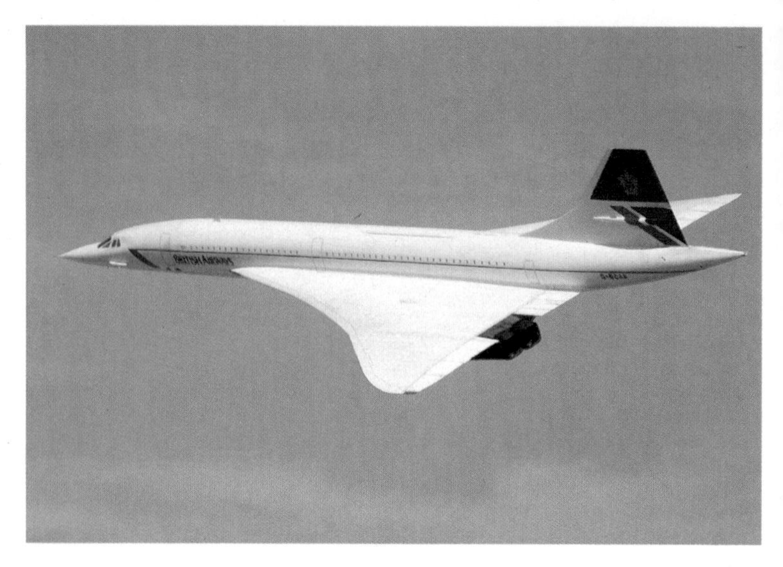

Above:
Concorde remains the world's only supersonic airliner in service.

Below:
Concorde with its nose drooped on finals to land.

Boeing 707/720

Four turbofan medium/long range transport

Basic data for Boeing 720B

Powerplant: Four Pratt & Whitney JT3D-1 of 17,000lb st
Span: 130ft 10in (39.88m)
Length: 136ft 9in (41.88m)
Max cruise: 608mph (978km/hr)
Passengers: 167 plus three/four crew

First aircraft flown: 15 July 1954 (Boeing 707-80)
Production: 967 Boeing 707 of which the main variants were the -320B and -320C (482 built); Boeing 720 (153 built). A total of 245 707/720s were in airline service early in 1988
Recent/current service with: Air Portugal, Air Zimbabwe, CAAC, Egyptair, Iran Air, Royal Jordanian, Nigeria Airways, Olympic, Pakistan International, Saudia, Tarom, Varig, Zambia Airways and many others
Recognition: Underwing mounted engines in four separate pods. Swept narrow chord wing, low-set circular, narrow body fuselage with the tailplane mounted either side of the tail cone. Tall, narrow fin and rudder, slightly swept, and a small ventral fin
Variants: The various models of the 707/720 differ mainly in fuselage length and powerplants. The original production 707-120 had a span of 130ft 10in (39.88mm) and a length of 144ft 6in (44.04m); the 120B was re-engined with turbofans; the 707-320 was larger with a span of 142ft 5in (43.41m) and length of 152ft 11in (46.61m); the turbofan-engined -320B had a span of 145ft 9in (44.42m) with the 152ft 11in (46.61m) fuselage, while the -320C was similar but featured a large, port-side forward fuselage freight door. The 720 had the original short-span wings and a shorter fuselage of 136ft 2in (41.50m); the 720B was a turbofan-engined variant

Below:
The Boeing 707 has a swept narrow chord wing carrying the four podded engines.

Above:
This cargo version of the Boeing 707 is operated by Tarom, the Rumanian national carrier. *Andrew March*

Below:
A Boeing 707-347C of Middle East Airlines.

Boeing 747

Four turbofan long range 'jumbo' airliner

Basic data for Boeing 747-200B

Powerplant: Four Pratt & Whitney JT9D-7R4 of 54,750lb st
Span: 195ft 8in (59.64m)
Length: 231ft 10in (70.7m)
Max cruise: 584mph (940km/hr)
Passengers: Up to 534 plus three crew

First aircraft flown: 9 February 1969
Production: Over 709 built by late 1988 with a further 170 ordered, including 100 of the new -400 variant
Recent/current service with: Most of the world's major airlines including Air Canada, Air France, Air India, All Nippon, American Airlines, Alitalia, British Airways, Flying Tiger, Iran Air, Japan Airlines, Korean Airlines, Pan American, Qantas, Singapore Airlines, Swissair, TWA and United
Recognition: Underwing mounted engines in four separate nacelles. Swept, low-set wing which narrows towards the tips. Oval, wide body fuselage with a distinctive raised fuselage forward of the wing, incorporating the cabin and flight deck. Tall, swept fin with a fuselage mounted tailplane below the rudder
Variants: There are 10 main variants of the 747, varying with the different powerplants installed, cabin configuration and weight specifications of the purchasing airlines. Most are hard to distinguish externally. Those most recognisable are the 747 Combi, 747F, 747SP and the 747-300 (stretched upper deck). The Boeing 747 Combi is a standard size 747 fitted with a large, port-side freight door. The 747F is fitted with an upward hinging nose for freight loading. This special all-cargo version does not have cabin windows. There are some 747 Combis fitted with the 'F' hinging nose. The 747SP has a 47ft (14.40m) shorter fuselage with a taller fin and rudder and new wing flaps. The raised forward fuselage remains giving a short 'dumpy' look to the aircraft. This 'special performance' version is operated by Pan Am, TWA, Qantas, Iran Air, South African Airways and other airlines. The -300 variant has a stretched upper deck, some 23ft 4in (7.11m) longer than the standard 747. The latest variant is the -400 which has a higher gross weight, more powerful engines, and extended wing-tips 15ft 4in (4.7m) with 'winglets'

Above:
This Boeing 747-233B is operated by Air Canada.

Below:
The shorter Boeing 747SP has a very tall fin and rudder.

Above:
The -300 Srs 747 has a longer fuselage and stretched upper deck.

Below:
The latest 747 is the Series 400.

British Aerospace BAe 146

Four turbofan short range airliner

Basic data for BAe 146-100

Powerplant: Four Avco Lycoming ALF502R-5 of 6,970lb st
Span: 85ft 5in (26.04m)
Length: 85ft 10in (26.16m)
Max cruise: 482mph (776km/hr)
Passengers: 93 plus two crew

First aircraft flown: 3 September 1981
Production: By 1988 a total of 108 built with 25 on order
Recent/current service with: Dan-Air; SATA Azores; Air Wisconsin, Air UK, American AL, Aspen AW, CAAC, Presidential AW, TNT United Express and US Air
Recognition: Underwing mounted engines in four nacelles. Slightly swept wings mounted on top of the fuselage, drooping towards the wing tips. Distinctive trailing edge wing fillets. The fuselage is circular in section with bulges on the lower side to accommodate the undercarriage. The rectangular fin and rudder is slightly swept with a T-tailplane mounted on the top of the fin. There are sideways opening airbrakes below the rudder
Variants: The initial BAe 146-100 was followed by the 8ft (2.44m) stretched -200 series. The fuselage was further stretched to produce the -300 series giving an overall length of 101ft 8in (30.99m). The QT (Quiet Trader) is offered as a freight version of the series 100/200 and features a large freight door. The international freight carrier TNT has an agreement to purchase 70 QTs

Below:
A BAe 146-200 aircraft operated by Air UK.

Ilyushin IL-62

Four turbofan long range airliner

Basic data for Ilyushin IL-62M

Powerplant: Four Soloviev D-30KU of 24,250lb st
Span: 141ft 9in (43.20m)
Length: 174ft 3½in (53.12m)
Max cruise: 560mph (901km/hr)
Passengers: 186 plus five crew

First aircraft flown: January 1963. Entered service in 1967
Production: Over 240 built for Aeroflot and Soviet bloc airlines
Recent/current service with: Aeroflot, CSA, Interflug, LOT, and Tarom, also CAAC and Cubana

Recognition: Rear fuselage side mounted engines in two double nacelles. Swept wings, low-set on narrow body, circular fuselage. T-tailplane mounted on top of the swept fin and rudder. Distinctive bullet fairing projecting forward of the fin
Variants: The improved IL-62M has more powerful engines than the original IL-62 and, like the IL-62MK which carries more passengers, is externally identical

Below:
The Ilyushin IL-62 has rear-fuselage-mounted engines and a T-tail.

Ilyushin IL-76

Four turbofan long range transport

Basic data for Ilyushin IL-76T

Powerplant: Four Soloviev D-30KP of 26,455lb st
Span: 165ft 8in (50.50m)
Length: 152ft 10½in (46.60m)
Max cruise: 497mph (800km/hr)
Payload: 88,185lb or 90 passengers and five crew

First aircraft flown: 25 March 1971
Production: Over 250 IL-76T and IL-76M built for civil airline use including more than 100 for Aeroflot
Recent/current service with: Aeroflot, Cubana, Iraqi Airways, Jamahiriya, and Syrian Arab Airlines
Recognition: Underwing mounted engines in four nacelles. Slightly swept wings mounted on top of the fuselage, drooping towards the wing tips. Circular fuselage with large bulges on either side and below the lower fuselage section for the undercarriage. Rectangular swept fin and rudder on the raised rear fuselage, with the swept T-tailplane mounted on top of the fin. A bullet fairing projects forward from the junction of the fin and tailplane. The aircraft nose had distinctive windows in the lower half with a circular bulge behind
Variants: The latest variant is the IL-76TD which has internal cabin improvements and marginally better performance

Below:
Iraqi Airways uses this IL-76MD for freight operations.

Above:
The glazed lower nose cone is apparent on this IL-76M of Soviet national carrier Aeroflot. *Andrew March*

Ilyushin IL-86

Four turbofan long range airliner

Basic data for Ilyushin IL-86

Powerplant: Four Kuznetsov NK-86 of 28,660lb st
Span: 157ft 8¼in (48.06m)
Length: 195ft 4in (59.54m)
Max cruise: 590mph (949km/hr)
Passengers: 350 plus three crew

First aircraft flown: 22 December 1976
Production: About 60 built by 1988 for airline use
Recent/current service with: Aeroflot

Recognition: Underwing mounted engines in four nacelles. Swept wings mounted on the lower section of the circular, wide-body fuselage. Swept fin and rudder with low-set swept tailplane mounted on the rear fuselage. Lower fuselage fairings beneath the wings
Variants: None to date

Below:
The IL-86 was the first Soviet wide-bodied airliner.

Above:
Aeroflot flies the IL-86 on long range routes.

McDonnell Douglas DC-8

Four turbofan long range transport

Basic data for DC-8 Series 50

Powerplant: Four Pratt & Whitney JT3D-1 of 17,000lb st
Span: 142ft 5in (43.41m)
Length: 150ft 6in (45.87m)
Max cruise: 595mph (958km/hr)
Passengers: 179 plus three crew

First aircraft flown: 30 May 1958 (DC-8-10); 20 December 1960 (DC-8-50)
Production: 556 of all variants, of which some 260 remain in service, mostly being Super 60/70 series
Recent/current service with: Many airlines including Air Canada, Emery Worldwide, Faucett, German Cargo, Japan Airlines, Scanair, Spantax, United, United Parcel Service and Zantop
Recognition: Underwing mounted engines in four pods. Swept wings mounted below the circular, narrow body fuselage. Tall, slightly swept fin and rudder with a low swept tailplane mounted on the rear fuselage. Super 60/70 series have an extremely long fuselage

Variants: The Series 10 through to 50 had differences of powerplant and performance but are externally all similar. The Series 55 had a large cargo door. Major changes came with the Super 60 series which had improved turbofan engines; the Series 61 had a considerably lengthened fuselage (36ft/11.18m) longer; the Series 62 had a modest increase in length (6ft 8in/2.03m longer), extended wing tips making the span 148ft 5in (45.24m) and modified engine pods; the Series 63 has the longer fuselage (187ft 5in/57.12m total) and the extended wings of the Series 62. The re-engined Series 60 and 70 have CFM56 turbofans which are shorter and of much greater diameter than the earlier powerplants

Below:
This SAS DC-8-63 has the longer fuselage characteristic of later versions of the DC-8.

Above:
The DC-8-71 has been re-engined with the distinctive large intake CFM-56 turbofans. *Neil Hargreaves*

Three-engined jet airliners

Boeing 727

Three turbofan medium range airliner

Basic data for Boeing 727-200

Powerplant: Three Pratt & Whitney JT8D-7 of 16,000lb st
Span: 108ft 0in (39.92m)
Length: 153ft 2in (49.69m)
Max cruise: 599mph (964km/hr)
Passengers: 189 plus three crew

First aircraft flown: 9 February 1963 (Series 100)
Production: A total of 1,832 built when production finished in August 1984, including 407 Series 100, 164 with large freight doors and 1,260 Series 200s and Advanced 200s
Recent/current service with: 1,700 were in service with many airlines worldwide including Air Atlantis, Air Canada, Air France, American Airlines, Ansett, Continental, Dan Air, Delta, Eastern, Federal Express, Iberia, Korean, Libyan Arab, Lufthansa, Mexicana, Northwest, Pan American, Taa, Tunis Air, United, Varig and Australian
Recognition: Three rear mounted engines, one on top of the fuselage at the base and forward of the fin, the other two in line either side of the rear fuselage. Low-set swept wings mid-way along the circular narrow-body fuselage. Swept fin and rudder with T-tailplane on top of the fin
Variants: The original 407 Series 100s were 10ft (3.05m) shorter than the Series 200 which was first flown on 27 July 1967 with more powerful engines and other improvements. A freight variant with a large cargo door (C) and a quick-change variant (QC) with the large door and palletised passenger seats have also been built. The Advanced 200 has further powerplant and internal refinements

Below:
Aviogenex of Yugoslavia operates this Boeing 727-2L8.

Above:
Pan Am operates Boeing 727s on European services from West Germany and Berlin.

Lockheed TriStar

Three turbofan long range airliner

Basic data for TriStar 100

Powerplant: Three Rolls-Royce RB211-22B of 42,000lb st
Span: 155ft 4in (47.35m)
Length: 177ft 8in (54.15m)
Max cruise: 575mph (925km/hr)
Passengers: 400 plus three crew

First aircraft flown: 17 November 1970
Production: 250 built (including a development aircraft) by 1983 when production ceased
Recent/current service with: 22 airlines worldwide including: Air Canada, Air Lanka, Air Portugal, Alia, All Nippon, British Airways, Cathay Pacific, Delta, Eastern, Gulf Air, LTU, Saudia and TWA
Recognition: Two engines in under-wing nacelles and one engine mounted on top of the fuselage forward of the swept fin, with the jet efflux below the rudder through the tail cone. Circular wide-body fuselage with low-set swept wings at mid-way point. Swept tailplane low-set either side of the rear fuselage below the fin
Variants: TriStar Series 1, 100 and 200 have powerplant and internal differences only. The major variant is the longer L-1011-500 (first flown 18 October 1978) which has extended wings to give a span of 164ft 4in (50.09m), 13ft 6in (4.12m) shorter fuselage (to 164ft 2in [50.04m]) and other internal modifications. The -500 can be recognised by its shorter fuselage, particularly forward of the wing, and in detail the reduction of port-side doors from four on the earlier models to three on this variant

Below:
The TriStar 1, here in British Airways colours, has two under-wing and one tail-mounted engine.

McDonnell Douglas DC-10/MD-11

Three turbofan long range airliner

Basic data for DC-10 Series 30

Powerplant: Three General Electric CF6-50C of 52,500lb st
Span: 165ft 4in (50.42m)
Length: 181ft 7in (55.35m)
Max cruise: 565mph (909km/hr)
Passengers: 380 plus three crew

First aircraft flown: 29 August 1970
Production: Over 440 by 1988 (including KC-10s for the USAF). The MD-11 followed the DC-10 in production from 1987
Recent/current service with: Nearly 50 major airlines worldwide including Air Afrique, American Airlines, Canadian Pacific, Continental, Finnair, Garuda, Iberia, Japan Airlines, KLM, Korean Airlines, Lufthansa, Mexicana, Martinair, Northwest, Sabena, Swissair, THY, United, UTA, Varig, Wardair and World Airways
Recognition: Two engines in under-wing nacelles close to the fuselage, one engine mounted on the fin above the fuselage with a straight-through exhaust pipe to the rear. A circular, wide-body fuselage with low-set, swept wings. Tailplane mid-set on the rear fuselage below the fin
Variants: The initial two variants, DC-10-10 and -15, were externally identical; the -30 had a 10ft (3.05m) extension to the wing giving a span of 165ft 4in (50.39m) while the re-engined (Pratt & Whitney JT9D) -40 is slightly longer at 182ft 3in (55.55m). The only DC-10 variant more easily recognised is the combi/cargo (CF) version with a large port-side forward freight loading door. The 405 seat MD-11 is a development with 18ft 6in (5.65m) longer fuselage, modified wing with tip winglets and a smaller tailplane. Powerplant will be 60,000lb st Pratt & Whitney PW 4360 or 61,500lb st GE CF-6-80C turbofans

Below:
This striking orange and silver colour scheme is on a DC-10-30 of CP Air.

Above:
The third engine on the DC-10 is mounted above the fuselage and exhausts directly to the rear.

Tupolev Tu-154

Three turbofan medium range airliner

Basic data for Tupolev Tu-154B

Powerplant: Three Kuznetsov NK-8-211 of 23,150lb st
Span: 123ft 2½in (37.55m)
Length: 157ft 2in (47.90m)
Max cruise: 601mph (967km/hr)
Passengers: 164 plus three crew

First aircraft flown: 4 October 1968
Production: Over 680 built by 1988
Recent/current service with: Over 500 in service with Aeroflot; also operated by Alyemda, Balkan Bulgarian, Cubana, LOT, Malev and Tarom
Recognition: Two engines mounted either side of the rear fuselage with the third engine on top of the rear fuselage forward of the fin, exhausting through tail cone. Very swept wings set below the circular, narrow body fuselage. Fairings for the undercarriage extend to the rear of the wings. A swept T-tailplane mounted on top of the fin and rudder with a bullet fairing projecting forward of the tailplane/fin intersection. The wings appear to droop towards the tips
Variants: The three original versions operated in the West, the Tu-154, Tu-154A and Tu-154B, have no significant external differences, the main changes being in powerplant and internal improvements. The latest Tu-154M with Soloviev D-30KU engines and modified tailplane and spoilers is in service with Aeroflot, Balkan, CAAC, Syrian and LOT. A cargo version, the Tu154S has a port-side forward freight door and is in airline service

Below:
This Soviet designed Tupolev 154 is operated by the Hungarian airline Malev.

Above:
The tail-mounted engines and highly swept tailplane are evident on this Balkan Bulgarian Airlines Tu-154B-2.

Yakovlev Yak-40

Three turbofan short range airliner

Basic data for Yakovlev Yak-40

Powerplant: Three Ivchenko Ai-25 of 3,300lb st
Span: 82ft 0in (24.99m)
Length: 66ft 9½in (20.36m)
Max cruise: 342mph (550km/hr)
Passengers: 32 plus two crew

First aircraft flown: 21 October 1966
Production: Approximately 1,000 built, with over 800 in airline service
Recent/current service with: Aeroflot, Air Guinee, Balkan Bulgarian, CSA, Cubana, Hang Khong Vietnam, Lao Aviation, Syrian Air and TAAG-Angola AL

Recognition: Rear mounted engines with two on the fuselage sides behind the wings and the third in line above the fuselage forward of the swept fin and rudder, exhausting through the tail cone. Straight wings mounted below the circular fuselage, with noticeable dihedral and drooping flaps. T-tailplane mounted on top of the fin; the rudder extends beyond the rear fuselage. The port side of the fuselage has only eight circular cabin windows with a single passenger door forward of these
Variants: None apparent

Below:
The short range Yak-40 was built for domestic routes.

Twin-engined jet airliners

Aerospatiale SE310 Caravelle

Twin turbojet medium range airliner

Basic data for Caravelle VI-R

Powerplant: Two Rolls-Royce Avon 532R of 12,600lb st
Span: 112ft 6in (34.29m)
Length: 105ft 0in (32.00m)
Max cruise: 525mph (845km/hr)
Passengers: 99 plus three crew

First aircraft flown: 25 May 1955 (Caravelle I)
Production: 282 built (I-20, IA-12, III-78, VI-N-53, VI-R-56, VII-1, 10A-1, 10B-22, 10R-20, 11R-6, 12-12)
Recent/current service: About 65 remain in use, mostly with charter airlines including Air Inter, Corse Air, CTA, Europe Air Service, Hispania, Kabo Air, Minerve, Sterling and Syrian Air
Recognition: Engines mounted on either side of the rear fuselage forward of the fin. Low-set, slightly swept wings with distinctive wing fences. Curved leading edge to the fin with a straight trailing edge to the rudder. Swept tailplane mounted on the fin above the fuselage with a large bullet fairing to the rear
Variants: There are nearly a dozen variants of the Caravelle. The initial I, IA and III are externally the same; the only visual difference of the VI-N is the noise suppressors fitted to the engines, while the VI-R had thrust reversers. The Caravelle 10B had a 3ft 4in (1.02m) longer fuselage, revised wing leading edge and modified flaps; the 11R had a front fuselage freight loading door; and the Super Caravelle 12 had a further 10ft (3.05m) fuselage stretch and was re-engined with Pratt & Whitney JT8D turbofans

Below:
The final version of the Caravelle was this Super 12.

Above:
A Caravelle VI-N operated by the Corsican airline Corse Air. *Andrew March*

Airbus A300

Twin turbofan medium range airliner

Basic data for Airbus A300B4-200

Powerplant: Two General Electric CF6-50C2 of 52,500lb st
Span: 147ft 1in (44.84m)
Length: 175ft 9in (53.57m)
Max cruise: 552mph (888km/hr)
Passengers: 345 plus three crew

First aircraft flown: 28 October 1972
Production: Over 286 delivered by 1988 out of 321 orders
Recent/current service with: Over 40 airlines worldwide, including; Air France, Air India, Air Inter, Air Jamaica, Alitalia, Eastern Airlines, Egyptair, Iberia, Indian Airlines, Iran Air, Lufthansa, Olympic, Pan Am, Thai International, Saudia and TOA
Recognition: Engines are carried in nacelles under and projecting forward of the wings. Circular wide-body fuselage, tapering towards the tail. Low-set, swept wings with five underwing fences/ flap fairings. Tall, swept fin and rudder and a swept tailplane set either side of the rear fuselage cone below the rudder
Variants: Differ mainly in powerplants and internal refinements. The A300C4 features a large fuselage side-loading door. First major difference comes with the A300B4-600 which has a modified (A310) rear fuselage with a longer parallel section and more sharply tapered tail cone. The A310 is a smaller sized development of the A300

Below:
This Airbus A300-B4 flies with the Spanish national airline Iberia.

Above:
Alitalia is one of many European airlines operating the Airbus A300.

Airbus A310

Twin turbofan long/medium range airliner

Basic data for Airbus A310-200

Powerplant: Two General Electric CF6-8A1 of 48,000lb st
Span: 144ft 0in (43.89m)
Length: 153ft 2in (46.66m)
Max cruise: 553mph (890km/hr)
Passengers: 280 plus three crew

First aircraft flown: 3 April 1982
Production: Over 160 ordered with over 132 in service by late 1988
Recent/current service with: Air Algerie, Air France, CAAC, Cyprus Airways, Kenya Airways, KLM, Kuwait AW, Lufthansa, Pan Am, Sabena, Singapore Airlines, Swissair THY and Wardair
Recognition: Underwing mounted engines in nacelles projecting forward of the wings. Circular wide-body fuselage tapering upwards towards the tail. Low-set, swept wings with four underwing fences/fairings. Tall, swept fin and rudder and a swept tailplane set either side of the rear fuselage cone below the rudder. The A310 is 22ft 10in (6.96m) shorter than the A300 and has a new wing of shorter span
Variants: The A310-200 and -220 differ only in powerplants while the -300 features wing-tip fences and interior refinements

Below:
Winglets identify this latest version of the A310, the 300 Series.

Above:
This A310 is operated by the Turkish airline THY.

Airbus A320

Twin turbofan medium range airliner

Basic data for Airbus A320

Powerplant: Two CFM 56-5 turbofans of 25,000lb st
Span: 111ft 3in (33.91m)
Length: 123ft 3in (37.57m)
Max cruise: 561mph (903km/hr)
Passengers: 179 maximum plus two crew

First aircraft flown: 22 February 1987
Production: 287 ordered by Air France, Air Inter, Air Malta, Adria, All Nippon, Ansett, Australian AL, British Airways (ex-BCAL order), Indian Airlines, Lufthansa, Northwest and Pan American. First delivery to Air France on 18 April 1988 and to British Airways on 20 April 1988
Recognition: Underwing mounted engines in nacelles protruding forward of the wings. Circular narrow body fuselage tapering upwards towards the tail. Low-set, swept wings of narrow chord with three tailing edge fairings. Tall, swept fin and rudder set forward of tail cone. Swept tailplane with dihedral

Below:
The A320 first entered service with British Airways in 1988. *Airbus*

Boeing 737

Twin turbofan medium range airliner

Basic data for Boeing 737-200

Powerplant: Two Pratt & Whitney JT8D-25A of 16,000lb st
Span: 93ft 0in (28.35m)
Length: 100ft 2in (30.53m)
Max cruise: 576mph (927km/hr)
Passengers: 130 plus two crew

First aircraft flown: 9 April 1967
Production: 2,212 ordered by late 1988 including 30 Series 100, 1,144 Series 200, 790 Series 300, 145 Series 400 and 133 Series 500. Approximately 1,600 Boeing 737s are in service
Recent/current service with: 120 airlines worldwide including Aer Lingus, Air Europe, Air Florida, Air France, All Nippon, Braathens, Britannia, British Airways, Canadian Airlines, Frontier, Indian Airlines, Lufthansa, Monarch, Orion, People Express, Piedmont, Sabena, Saudia, South African Airways, Southwest, United Airlines, Varig, Vasp, Western and Wien Air
Recognition: Engines mounted directly under the swept wings. Tubby, circular fuselage with wings set in the lower section Tall, angular, slightly swept fin and rudder with the swept tailplane set on the rear fuselage at the base of the rudder
Variants: Series 100 had a 6ft (1.83m) shorter fuselage, but only 30 were built before the larger Series 200 entered production. The -200C has a large freight door on the port side of the forward fuselage. Series 300 which entered production in 1984 has a longer fuselage (109ft 7in/33.40m), slightly increased span wings (94ft 9in/28.90m), an extended dorsal fin, new CFM56-3 powerplants with more circular section nacelles projecting forward of the wing and other minor changes. Entering service in 1988 the Series 400 is a 150-plus seater with the fuselage further extended to 119ft 7in (36.30m). The Series 500 is a 132 seat development with CFM 56-3 turbofans and a shorter fuselage of 101ft 9in length making use of latest technology construction and flight systems

Below:
A shamrock on the fin identifies this Boeing 737-248 as being one of the Aer Lingus fleet.

Above:
British Midland operates this new 300 series Boeing 737.

Above:
The B737-300 is powered by CFM56 turbofans.

Boeing 757

Twin turbofan medium range airliner

Basic data for Boeing 757-200

Powerplant: Two Rolls-Royce RB211-535E4 of 40,100lb st
Span: 124ft 10in (38.05m)
Length: 155ft 3in (47.32m)
Max cruise: 569mph (916km/hr)
Passengers: 239 plus three crew

First aircraft flown: 19 February 1982
Production: Over 360 ordered and 193 delivered by late 1988
Recent/current service with: Air Europe, Air 2000, American, British Airways, Delta, Eastern Airlines, Monarch, Northwest, Royal Air Maroc, Royal Brunei, Singapore Airlines and United Parcel Service

Recognition: Engines in nacelles under the wings. A very long, circular, narrow-body fuselage with swept, low-set wings at the mid-way point. A pronounced lower fuselage bulge for the undercarriage fairing. A tall, swept fin and rudder with a swept tailplane on either side of the rear fuselage below the fin
Variants: The Boeing 757PF (Package Freighter) has a forward cargo door and a windowless fuselage. The first aircraft of an order for 20 was delivered to United Parcel Service in September 1987

Below:
Boeing 757s have long narrow fuselages.

Boeing 767

Twin turbofan medium range airliner

Basic data for Boeing 767-200

Powerplant: Two Pratt & Whitney JT9D-7R4D of 48,000lb st
Span: 156ft 1in (47.57m)
Length: 159ft 2in (48.51m)
Max cruise: 582mph (937km/hr)
Passengers: 220 plus two crew

First aircraft flown: 26 September 1981
Production: Over 307 ordered with 236 in service
Recent/current service with: 24 airlines including; American Airlines, Air Canada, Britannia, Delta, Egyptair, El Al, Kuwait Airways, Transbrasil, TWA and United Airlines

Recognition: Turbofans mounted under the swept wings. Circular fuselage with the wings set in the lower section, mid-way between the nose and tail. Very tall swept fin with the tailplane set on the rear fuselage at the base of the rudder
Variants: The stretched 767-300, which has a 21ft 1in (6.43m) longer fuselage, entered service with Japan Airlines in November 1986. British Airways is the launch customer (11) for an RB-211-engined version of the 767-300ER

Below:
This colourful Boeing 767-266ER flies with EgyptAir. *Andrew March*

Above:
This twin-jet Boeing 767-233ER is flown by Air Canada on trans-Atlantic flights.

British Aerospace BAC 1-11

Twin turbofan short range airliner

Basic data for BAC 1-11 Series 500

Powerplant: Two Rolls-Royce Spey 512 DW of 12,550lb st
Span: 93ft 6in (28.50m)
Length: 107ft 0in (32.64m)
Max cruise: 541mph (871km/hr)
Passengers: 119 plus two crew

First aircraft flown: 20 August 1963
Production: 230 in UK (56 Series 200, nine Series 300, 69 Series 400, nine Series 475, 87 Series 500), Series 560 being built by CNIAR in Rumania as the Rombac 1-11
Recent/current service with: Adria, Aer Lingus, Air Malawi, Austral, British Airways, British Island Airways, Cyprus Airways, Dan Air, Florida Express, Okada Air, Philippine Airlines, Ryanair, Tarom and US Air
Recognition: Engines mounted either side of the rear fuselage forward of the tail unit. Slightly swept wings set in the lower section of the circular narrow body fuselage. T-tailplane mounted on top of the swept fin and rudder. Auxiliary power unit (APU) in the tailcone
Variants: Series 200, 300 and 400 had a short fuselage (93ft 6in/ 28.50m) and shorter span wings (88ft 8in/26.95m) and were broadly the same externally in appearance. The larger Series 500 had a 13ft 6in (4.14m) longer fuselage and a 5ft (1.55m) extended wing span. The Series 475 had the short fuselage of the Series 400 and the bigger wings of the Series 500, together with a modified undercarriage with low pressure tyres and larger wheels. Some aircraft have been retrofitted with a large forward freight door; others have had power plant 'hush kits'. Some Series 200/400 aircraft are being re-engined with the Rolls-Royce Tay turbofan. It is possible that new Rombac Series 500s might also be fitted with this powerplant

Below:
BAC 1-11 Srs 208ALs operated by Aer Lingus have hush-kits fitted to the Rolls-Royce Spey engines.

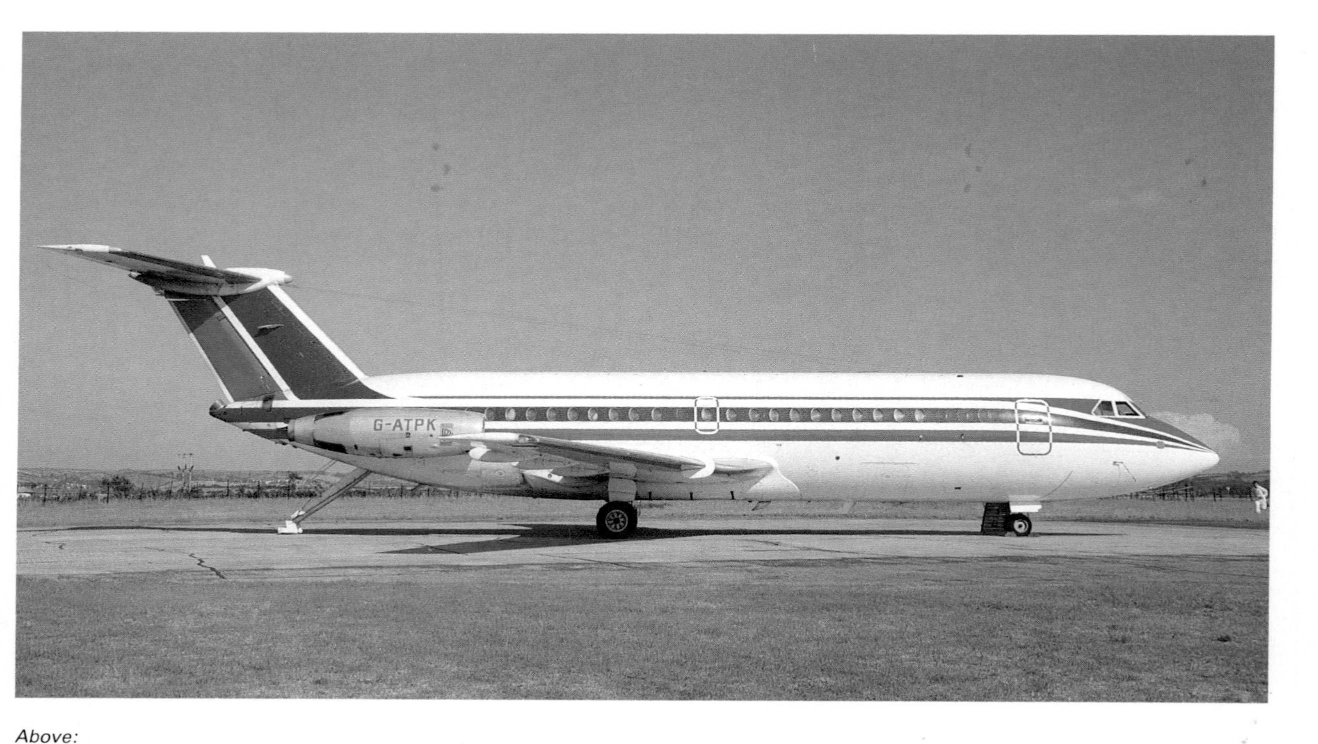

Above:
A BAC 1-11 Srs 301 operated for VIP passenger charters. *Andrew March*

Above:
The BAC 1-11 Srs 509 has a longer passenger cabin and increased wingspan.

Above:
BAC 1-11 production continues in Rumania, with CNIAR building the Rombac 1-11 Srs 561RC, in service with Tarom and Ryanair.
Andrew March

Dassault-Breguet Mercure

Twin turbofan short range transport

Basic data for Mercure 100

Powerplant: Two Pratt & Whitney JT8D-15 of 15,500lb st
Span: 100ft 3in (30.56m)
Length: 114ft 4in (34.85m)
Max cruise: 579mph (932km/hr)
Passengers: 162 plus two crew

First aircraft flown: 28 May 1971
Production: Only 12 aircraft built
Recent/current service with: Air Inter (11 aircraft in 1988)
Recognition: Engine nacelles mounted under the wings close inboard. Swept wings set on the lower section of the circular fuselage which sweeps up towards the rear and extends beyond the tail unit. Tall fin and rudder with a tapered, rounded top. Tailplane has marked dihedral and is set on top of the fuselage below the rudder
Variants: None

Below:
Air Inter operates the only production Mercures.

Fokker F28 Fellowship

Twin turbofan short range airliner

Basic data for Fokker F28-4000 Fellowship

Powerplant: Two Rolls-Royce RB183-2 Mk 555-15P of 9,900lb st
Span: 82ft 3in (25.07m)
Length: 97ft 2in (29.62m)
Max cruise: 524mph (843km/hr)
Passengers: 85 plus two crew

First aircraft flown: 9 May 1967
Production: 241 (production complete)
Recent/current service with: 34 operators including Aero Peru, Braathens, Ghana Airways, Garuda, LADE, Linjeflyg, NLM Cityhopper, Piedmont, TA Transregional and THY
Recognition: Engines mounted on the sides of the rear fuselage forward of the tail. Low-set wings in the lower section of the circular, short fuselage. Swept fin with a rounded top and a dorsal fillet. Swept tailplane set at the top of the fin with a rounded section projecting above the tailplane. Sideways opening airbrakes below the rudder
Variants: F28 Mks 1000, 3000 and 5000 have shorter fuselages than the Mks 2000, 4000 and 6000; Mks 3000 to 6000 have extended wings, the Mks 5000 and 6000 with leading edge slats

Mk 1000 span 77ft 4in (23.57m), length 89ft 11in (27.41m)
Mk 2000 span 77ft 4in (23.57m), length 97ft 2in (29.62m)
Mk 3000 span 82ft 3in (25.07m), length 89ft 11in (27.41m)
Mk 4000 span 82ft 3in (25.07m), length 97ft 2in (29.62m)
Mk 5000 span 82ft 3in (25.07m), length 89ft 11in (27.41m)
Mk 6000 span 82ft 3in (25.07m), length 97ft 2in (29.62m)

There are also a limited number of cargo (C) versions with a large port-side loading door

Below:
A Fokker F28 Fellowship 1000 flown by TAT for Air France.

Fokker 100

Twin turbofan short/medium range airliner

Basic data for Fokker 100

Powerplant: Two Rolls-Royce Tay 620-15 turbofans of 13,850lb st
Span: 92ft 1½in (28.08m)
Length: 116ft 7in (35.53m)
Max cruise: 497mph (800km/hr)
Passengers: 107 plus two crew

First aircraft flown: 30 November 1986
Production/service with: KLM and Swissair from 1988. Ordered also by US Air (20) and Air Ivoire (1)
Recognition: Engines mounted on the sides of the rear fuselage aft of the wing. Swept tailplane set at the top of the fin which has a rounded top and dorsal fillet. Much longer fuselage than the F28, circular in section. Low-set wings of increased span and modified tips

Below:
Delivery of Rolls-Royce Tay-powered Fokker 100s is now under way.

McDonnell Douglas DC-9 Series 10-50

Twin turbofan short/medium range airliner

Basic data for DC-9 Series 30

Powerplant: Two Pratt & Whitney JT8D-9 of 14,500lb st
Span: 93ft 5in (28.47m)
Length: 119ft 4in (36.37m)
Max cruise: 579mph (932km/hr)
Passengers: 115 plus two crew

First aircraft flown: 25 February 1965 (Series 10)
Production: A total of 976 DC-9s (up to the Series 50) were built, of which 850 were still in airline service by early 1988
Recent/current service with: Aeromexico, Air Canada, Alitalia, Australian, Aviaco, British Midland, Continental, Delta, Eastern, Finnair, Garuda, Iberia, Adria Airways, JAT, KLM, Northwest, SAS, Spantax, TriStar, US Air and many others world-wide
Recognition: Engines mounted on the sides of the rear fuselage forward of the fin. Low set, swept wings which taper towards the wing tips. Narrow body, circular fuselage with a rounded nose and pointed tail. Angular swept fin and rudder with a swept T-tailplane mounted near the top. A small rounded fin extension above the tailplane with a bullet fairing to the rear
Variants: The DC-9 has been extended in length and wing span progressively from the Series 10 to the current production MD-80 series. There have also been DC-9C convertible and DC-9F freight variants, with large forward freight doors

DC-9-10 span 89ft 5in (27.25m), length 104ft 5in (31.83m)
DC-9-20 span 93ft 5in (28.47m), length 104ft 5in (31.83m)
DC-9-30 span 93ft 5in (28.47m), length 119ft 4in (36.37m)
DC-9-40 span 93ft 5in (28.47m), length 125ft 8in (38.30m)
DC-9-50 span 93ft 5in (28.47m), length 133ft 7in (40.72m)

The different models also have changes in powerplants, all-up-weights and internal improvements, etc

Below:
The smallest of the DC-9s is the Series 15, shown here in the colours of British Midland Airways.

Above:
Italian charter airline ATI uses the DC-9-32.

McDonnell Douglas MD-80 series

Twin turbofan short/medium range airliner

Basic data for MD-81

Powerplant: Two Pratt & Whitney JT8D-209 turbofans of 19,250lb st
Span: 107ft 10in (32.87m)
Length: 147ft 11in (45.08m)
Max cruise: 574mph (924km/hr) at 27,000ft
Passengers: maximum 172 plus two crew

First aircraft flown: 18 October 1979 (MD-80)
Production: 528 built out of a total order book of over 750
Recent/current service with: 37 airlines including Adria, American, Alitalia, Austrian, Continental, Delta, CAAC, Finnair, SAS, Swissair and US Air
Recognition: Engines mounted on the sides of the rear fuselage forward of the swept fin and rudder. Low-set swept wings which taper towards the tips. Non-swept inboard trailing edges and 10in taller fin. Narrow body, circular fuselage considerably extended forward of the wings compared with the earlier DC-9 series
Variants: Originally known as the Super 80, the MD-80 series has these sub-types:

MD-81 as above
MD-82 more powerful JT8D-217 turbofans of 20,850lb st
MD-83 higher all up weight and payload/range
MD-87 shorter fuselage length 130ft 6in (39.73m), JT8D-217 turbofans (20,859lb st) and carries 130 passengers
MD-88 longer fuselage length 147ft 11in (45.08m) and JT8D-217 turbofans (20,850lb st)

Below:
The MD81 development of the DC-9 has a stretched fuselage.

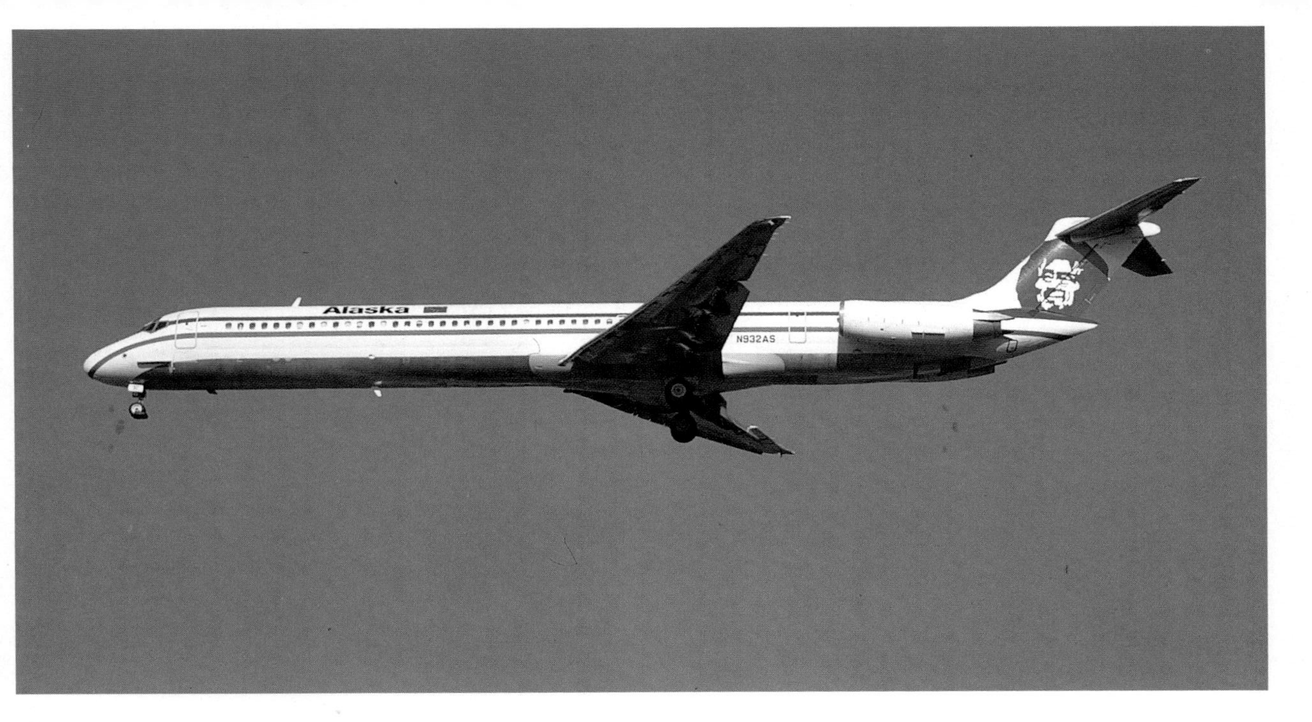

Above:
The smiling face of a friendly Eskimo looks down from the fin of this Alaska Airlines MD-83. *Neil Hargreaves*

Tupolev Tu-134

Twin turbofan short/medium range airliner

Basic data for Tupolev Tu-134A

Powerplant: Two Soloviev D-30-II of 14,990lb st
Span: 95ft 2in (29.01m)
Length: 121ft 7in (37.06m)
Max cruise: 550mph (885km/hr)
Passengers: 86 plus three crew

First aircraft flown: 1962
Production: 700 built with over 650 in airline service
Recent/current service with: Aeroflot, Aviogenex, Balkan Bulgarian, CSA, Interflug, LOT, Malev and Syrian Air
Recognition: Engines mounted high on the sides of the rear fuselage below the dorsal fin. Well swept wings mounted on the bottom of the slim, circular fuselage, with large undercarriage fairings extending from the trailing edges of the wings. A broad, slightly swept fin and rudder with the very swept tailplane mounted on top. Bullet fairings project forward and rear of the fin/tailplane junction
Variants: The original Tu-134 had a 6ft 11in (2.11m) shorter fuselage than the major production version, the Tu-134A

Below:
The Tupolev Tu-134A is used widely by Eastern bloc countries including the Polish airline LOT.

Above:
Balkan Bulgarian Airlines operates the twin-engined Tu-134A. *Daniel March*

Four-engined propeller airliners

Canadair CL-44

Four turboprop long range transport

Basic data for Canadair CL-44D-4

Powerplant: Four Rolls-Royce Tyne 515/10 of 5,730shp
Span: 142ft 3½in (43.37m)
Length: 136ft 10¾in (41.73m)
Max cruise: 392mph (631km/hr)
Passengers: 189 plus three crew; payload 63,272lb

First aircraft flown: 15 November 1959
Production: 39 built, mainly for civil use; 13 still in service early 1988
Recent/current service with: Aer Turas, Bayu Indonesia, HeavyLift, Katale AT, Volcanair and Wrangler
Recognition: Four turboprops set in the wings, which have a swept leading edge. Wings mounted on the lower fuselage at mid-point between the nose and tail. A long, slender fuselage and a large, broad fin and rudder. The tailplane is low-set on the rear fuselage. A large front, port-side, cargo door. The rear fuselage and tail are hinged, opening sideways below the dorsal fin
Variants: Developed from the Bristol Britannia which it resembles, there are four main variants of the CL-44. The CL-44D-6 is the ex-Canadian Armed Forces CC-106 Yukon; it does not have a swing-tail but does have two large port-side freight loading doors. The CL-44D-4 has the sideways-hinging rear fuselage. The CL-44J has an extended fuselage (151ft 9¾in/46.27m) with a hinged tail. Conroy conversion CL-440 has a massive upper fuselage bubble and a hinged tail

Below:
This CL-44D-4 has a sideways hinging rear fuselage allowing outsize cargo to be easily loaded.

Above:
HeavyLift operates the sole CL-440 with its massive upper fuselage bubble.

De Havilland Canada Dash 7

Four turboprop STOL short range airliner

Basic data for DHC Dash 7-100

Powerplant: Four Pratt & Whitney PT6A-50 of 1,120shp
Span: 93ft 0in (28.35m)
Length: 80ft 8in (24.59m)
Max cruise: 265mph (426km/hr)
Passengers: 56 plus two crew

First aircraft flown: 27 March 1975
Production: 105 built by 1988 and in service with 26 airlines
Recent/current service with: Air BC, Arkia, Atlantic Southeast, Brymon Airways, Emirates Air Services, Eurocity, Greenlandair, Hawaiian Airlines, Henson Airlines, Adria Airways, Maersk Air, Pan Am Express, Tyrolean Airways, Wideroe, Yemen Airways and others
Recognition: Four turboprops, each with five-blade propellers, mounted below the straight wing, which has distinctive trailing flaps. The circular fuselage is set below the wing, tapering upwards towards the tail. Very large swept fin and rudder with a dorsal extension; a straight tailplane is mounted on top of the fin
Variants: The only version produced to date is described above

Below:
The Dash 7 is the only airliner operating into London City STOLport.

Above:
London City Airways operates this Dash 7 on European routes.

Douglas DC-4/DC-6/DC-7

Four piston engined medium range transport

Basic data for Douglas DC-4

Powerplant: Four Pratt & Whitney R-2000-25D-13G of 1,450hp
Span: 117ft 6in (35.81m)
Length: 93ft 5in (28.47m)
Max cruise: 204mph (328km/hr)
Passengers: 86 plus three crew

First aircraft flown: 21 June 1938 (DC-4E); 14 February 1942 (C-54)
Production: Over 2,350 of the DC-4, 6, 7 series built including military versions
Recent/current service with: Many small airlines in North and South America including Aerial Transit, American Air Freight, Aesa Airlines, Conifair, Connor Airlines, Millardair, Northern AC, Satena, Trans-Air Link. Also Air Atlantique and Instone in the UK
Recognition: Four piston engines mounted on the tapered and dihedralled wings, which are set in the lower section of the circular fuselage. Rear, top fuselage tapers downwards; the straight fin and rudder has a curved top edge. The low-set tailplane is located on either side of the tail cone below the fin and rudder. Oval cabin windows and a large freight door on the port side of the converted ex-military aircraft
Variants: The Canadair C-54M was a variant built in Canada and powered by Rolls-Royce Merlin engines — none remains in airline service. The Carvair was a DC-4 modified as a car-ferry/freighter, with a raised cockpit and distinctive bulbous, opening nose and a modified fin and rudder. The DC-6 was a pressurised, longer development of the DC-4, with more powerful engines and other internal changes. The DC-7 was further increased in performance and internal refinement

DC-6 span 117ft 6in (35.81m), length 100ft 7in (30.66m)
DC-6A span 117ft 6in (35.81m), length 105ft 7in (32.18m), and modified, more angular fin and rudder, with fore and aft freight doors
DC-6B span 117ft 6in (35.81m), length 106ft 8in (32.51m), with nose weather radar. Fitted as passenger transport
DC-6C — modified DC-6A aircraft to passenger configuration
DC-7 span 117ft 6in (35.81m), length 108ft 11in (33.20m), more powerful engines, and modified forward fuselage
DC-7B span 117ft 6in (35.81m), length 108ft 11in (33.20m)
DC-7C span 127ft 6in (38.86m), length 112ft 3in (34.21m), long range development with increased span and fuselage length
DC-7F span 127ft 6in (38.86m), length 112ft 3in (34.21m), freighter conversion with port-side front cargo door and an extra large rear fuselage loading door

Above:
The piston-engined Douglas DC-4 continues to serve in small numbers, mainly in North and South America.

Above:
Air Atlantique uses this DC-6B for ad-hoc cargo flights. *Roger Wright*

Below:
A newly restored DC-6 waiting to re-enter service.

Ilyushin IL-18

Four turboprop medium range airliner

Basic data for Ilyushin IL-18E

Powerplant: Four Ivchenko AI-20M of 4,250eshp
Span: 122ft 8½in (37.40m)
Length: 117ft 9in (35.89m)
Max cruise: 419mph (674km/hr)
Passengers: 110 plus five crew

First aircraft flown: 4 July 1957
Production: 656 built, mainly for Aeroflot, about 200 remain in service
Recent/current service with: Aeroflot, Air Guinee, Balkan Bulgarian, CAAC, CSA, Cubana, Interflug, LOT, Malev and Tarom

Recognition: Four turboprops mounted above the straight wing which is set in the lower section of the long, circular fuselage. Tall, slender fin and rudder with a small dorsal extension. Straight tailplane set on either side of the fuselage below the rudder
Variants: Little external difference between the IL-18D, IL-18E and IL-18V production aircraft, which have revised powerplants and internal layouts

Below:
A Soviet-built IL-18D operated by Tarom.

Lockheed Electra

Four turboprop medium range transport

Basic data for L-188A Electra

Powerplant: Four Allison 501-D13 of 3,750eshp
Span: 99ft 0in (30.18m)
Length: 104ft 6in (31.85m)
Max cruise: 405mph (652km/hr)
Passengers: 98 plus three crew

First aircraft flown: 6 December 1957
Production: 170 built with about 76 in airline service in early 1988
Recent/current service with: Evergreen International, Falcon Cargo, Fred Olsen, Galaxy, Interstate Airlines, NWT Air, Mandala Airlines, Reeve Aleutian, Spirit of America Airways, TPI International Airways, Tramaco Transafrik, Varig and Zantop, International Airlines
Recognition: Four turboprops widely spaced on the broad chord, short span, square-tipped wings. Circular fuselage mounted above the wing. Shaped fin and rudder with a dorsal extension. Tailplane set on top of the rear fuselage below the fin and rudder
Variants: No external differences between the L-188A and L-188C. Most Electras remaining in service have been converted for freight operations with one or more large cargo loading doors on the port side of the fuselage

Below:
The distinctive profile of the turboprop Lockheed Electra.

Lockheed L-100 Hercules

Four turboprop long range transport

Basic data for L-100-30 Hercules

Powerplant: Four Allison 501-D22A of 4,680eshp
Span: 132ft 7in (40.41m)
Length: 112ft 9in (34.37m)
Max cruise: 363mph (584km/hr)
Passengers: 91 plus three crew; payload 50,738lb

First aircraft flown: 23 August 1954 (military YC-130); 21 April 1964 (L-382 airline variant)
Production: Only 104 civilian aircraft built from over 1,800 C-130s by 1988
Recent/current service with: Africargo, Air Gabon, CAAC, Markair, Merpati, Safair, Southern Air Transport, SF Air, Pelita & Uganda Airlines
Recognition: Four turboprops located under high-set straight wing. A circular fuselage with a distinctive nose radome, undercarriage fairings and up-swept tail for rear cargo loading. Tall, shaped fin and rudder with a small dorsal extension forward of the fin. Tailplane at the extremity of the fuselage at the base of the fin and rudder. Few cabin windows
Variants: The original L-382 was similar to the C-130E Hercules, Subsequent civil variants have had longer fuselages; L-100-20 was 8ft 4in (2.54m) longer and the L-100-30 was 20ft (6.10m) longer. Over 40 variants of the civil and military Hercules have been produced and are in service in 56 countries

Below:
One of a number of Lockheed L-100 Hercules in use by cargo airlines.

Shorts Belfast

Four turboprop medium range cargo transport

Basic data for Belfast

Powerplant: Four Rolls-Royce Tyne RTy 12 of 5,730shp
Span: 158ft 10in (48.41m)
Length: 136ft 5in (41.58m)
Max cruise: 325mph (566km/hr)
Passengers: 19 plus four crew; payload 75,000lb

First aircraft flown: 5 January 1964
Production: 10 built for the RAF; three converted for civil use
Recent/current service with: HeavyLift Cargo Airline

Recognition: Four turboprops mounted below the wing which has a swept leading edge and is set above the circular fuselage. Large undercarriage fairings on the lower fuselage section below and forward of the wing. Swept-up rear fuselage with a tall fin and rudder. Tailplane mounted on the fuselage at the base of the fin. A rear fuselage loading ramp
Variants: None

Below:
Three ex-RAF Belfasts are flown by HeavyLift for transporting outsize freight loads. *Andrew March*

Vickers Viscount

Four turboprop short range airliner

Basic data for Viscount 810

Powerplant: Four Rolls-Royce Dart 525s of 2,100eshp
Span: 93ft 8½in (28.56m)
Length: 85ft 8in (26.11m)
Max cruise: 385mph (620km/hr)
Passengers: 69 plus three crew

First aircraft flown: 16 July 1948 (Viscount 630 prototype)
Production: 440 built of which only 40 remained in service in 1988
Recent/current service with: Air Zimbabwe, Bouraq Indonesia, British Air Ferries, GB Airways, Guernsey Airlines, La Canarias, Mandala Airlines and Nusantra Airlines
Recognition: Four slender turboprops mounted on the wing leading edge projecting well forward. Low-set wings at the bottom of the oval fuselage which tapers towards the nose and tail. Shaped fin and rudder with a forward dorsal extension. Tailplane with marked dihedral set at either side of the rear fuselage below the fin and rudder. Large oval cabin windows
Variants: The original production 700 series Viscounts had a fuselage 81ft 10in (24.94m) long, while the improved 800/810 series had a 3ft 10in (1.17m) extension, mainly forward of the wing. Other changes included improved powerplants and cabin modifications

Below:
The Rolls-Royce Dart-engined Viscount continues in limited service in the UK.

Three-engined propeller airliners

Pilatus Britten-Norman Trislander

Three piston-engined short range commuter transport

Basic data for Trislander Mk III-2

Powerplant: Three Avco Lycoming 0-540-E4C5 of 260hp
Span: 53ft 0in (16.15m)
Length: 49ft 3in (15.01m)
Max cruise: 166mph (267km/hr)
Passengers: 17 plus one crew

First aircraft flown: 11 September 1970
Production: 73 built by 1982 when assembly switched to International Aviation Corporation, Miami
Recent/current service with: Aurigny Air Services, Loganair, XP Express and several US commuter airlines
Recognition: Unique three-engined layout. Two engines carried below the straight, high-set wings, with the third engine mounted on and projecting forward of the fin. Conical camber extended wing tips; fixed tricycle undercarriage, with the main wheels below the engines. A rectangular-shaped fuselage with an extended, circular, pointed nose. Broad angular fin rising above the engine and becoming taller rearwards. The tailplane is mounted either side of the rear engine above the rudder
Variants: The original Trislander had a short nose similar to the Islander. The Mk III-2 featured a 3ft 6in (1.07m) fuselage extension

Below:
The distinctive Trislander is flown by Aurigny on its Channel Island routes.
Andrew March

Above:
The 17-seat Trislander is the only three piston-engined airliner flying scheduled services.

Twin-engined propeller airliners

Aerospatiale Nord 262/Mohawk 298

Twin turboprop short range commuter airliner

Basic data for Nord 262A

Powerplant: Two Turbomeca Bastan VIC turboprops of 1,065shp
Span: 71ft 10in (21.89m)
Length: 63ft 3in (19.28m)
Max cruise: 233mph (375km/hr)
Passengers: 29 plus two crew

First aircraft flown: 24 December 1962
Production: Total of 110 built (N262A-67, N262B — four, N262C/D Fregate — 39)
Recent/current service with: Air Limousin, Air Littoral, CAdu Languedoc, Cimber Air, Pocono Airways and Pan Am Express
Recognition: Twin turboprops mounted forward of the straight wing which is set on top of the short, circular fuselage. Undercarriage fairings are located either side of the lower fuselage below the wings. The fin and rudder is rectangular with a dorsal fillet extending to above the rearmost fuselage window. A low-set tailplane is positioned either side of the tail cone
Variants: The first four production aircraft designated N262B, similar externally to the main production N262A. The N262C/D had a 2ft 3¾in (0.70m) increase in wing-span and uprated engines. Named the Fregate it was produced for the French Air Force and Navy. In the USA nine N262As were converted to Mohawk 298s by re-engining with Pratt & Whitney PT6A-45 turboprops

Below:
The French-designed Nord 262A is a short-range Astazou-powered regional airliner.

Aerospatiale — Aeritalia ATR42

Twin turboprop regional airliner

Basic data for ATR42

Powerplant: Two Pratt & Whitney Canada PW120-2 turboprops of 1,800shp
Span: 80ft 7½in (24.57m)
Length: 74ft 5¾in (22.70m)
Max cruise: 307mph (495km/hr) at 20,000ft
Passengers: 42 plus two crew

First aircraft flown: 16 August 1984
Production: 117 ordered by early 1988
Recent/current service with: 54 in service including Air Littoral, ATI, BritAir, Camber Air and other European commuter airlines; Command Airways, Simmons Airlines and Ransome in USA, Air Pacific and Air Tahiti

Recognition: High set straight wing with slim engines projecting forward and below the wing close inboard. Circular section fuselage with large undercarriage fairings under the centre section. Distinctive, large, slightly swept fin and rudder with two angle changes on the forward edge. Straight tailplane set near to the top of the fin
Variants: Two basic airliners ATR42-200 and -300, the latter with increased payload/range. ATR42-F is a cargo version with larger port-side door; ATR42-R has a rear fuselage loading ramp. Stretched ATR72 is due to enter service in 1989

Below:
Appropriately a pair of tulips adorn the fin of this Holland Aerolines ATR42.
Neil Hargreaves

Above:
The ATR42 has a distinctive high-wing and curiously shaped fin. *Neil Hargreaves*

Antonov An-24/26/30/32

Twin turboprop short range transport

Basic data for An-24V Series II

Powerplant: Two Ivchenko AI-24A turboprops of 2,530ehp
Span: 95ft 9½in (29.20m)
Length: 77ft 2½in (23.53m)
Max cruise: 310mph (499km/hr)
Passengers: 50 plus three crew

First aircraft flown: April 1960
Production: Over 1,100 An-24s built of which some 860 are in airline service
Recent/current service with: Aeroflot, Air Guinee, Air Mongol, Balkan, CAAC, Cubana, Iraqi Airways, Lao Aviation, LOT and Tarom
Recognition: Large twin turboprops mounted below and extending in front of and behind the wings, which taper sharply outboard of the engines. The circular section fuselage is set under the wings. A tall fin and rudder with a forward dorsal extension. The tailplane is mounted below the rudder on the fuselage. There is a ventral fin below the tail
Variants: The An-24T is an all-cargo version; the An-26 has a rear-loading ramp for military service; the An-30 is a developed aerial survey version with a new front fuselage; the An-32 is a more powerful version of the An-26

Below:
Loading of the Antonov An-26 is via a rear loading ramp enclosed by clam-shell doors. *Andrew March*

Beechcraft C99

Twin turboprop commuter airliner

Basic data for Commuter C99

Powerplant: Two Pratt & Whitney PT6A-36 turboprops of 715shp
Span: 45ft 10½in (13.98m)
Length: 44ft 6¾in (13.58m)
Max cruise: 249mph (401km/hr)
Passengers: 15 plus one/two crew

First aircraft flown: December 1965 (stretched Queen Air); 20 June 1980 (C99)
Production: Some 60 C99s have been delivered. 164 early Model 99, 99A and B99 were produced up to 1975
Recent/current service with: Commuter airlines in the USA and Europe

Recognition: Twin turboprops mounted on and well forward of low-set wing. A slab-sided fuselage with large entry doors aft of the port wing. Large rectangular cabin windows, a long pointed nose which dips downwards from the cockpit. A swept fin and rudder with both dorsal and ventral fillets. It has a swept tailplane set either side of the rear fuselage cone
Variants: Externally the Model 99, 99A and C99 are very similar in appearance

Below:
The Beech 99 remains in limited use.

Beech 1900L Super King Air

Twin turboprop commuter airliner

Basic data for Beech 1900C

Powerplant: Two Pratt & Whitney PT6A-65B turboprops of 1,100shp
Span: 54ft 6in (16.61m)
Length: 57ft 10in (17.63m)
Max cruise: 295mph (472km/hr)
Passengers: 19 plus one crew

First aircraft flown: 3 September 1982
Production: 60 delivered by the end of 1987
Recent/current service with: Commuter airlines in the USA
Recognition: Twin turboprops mounted on a low wing. Slab-sided fuselage with small circular cabin windows. Stabilon horizontal surfaces mounted on fuselage, forward of the tailplane. A swept fin and rudder with both dorsal and vertical fillets. High set swept tailplane with 'tail-ets' under the tailplane tips
Variants: 1900 airliner, King Air Exec-liner executive version and 1900C cargo airliner with large freight door aft of the wings. The Beech 200/300 Super King Air with shorter fuselage (without stabilons and tail-ets) is in use for airline and executive service. A specialised airliner version, the Beech 1300 with a 44cu ft belly cargo pack was ordered by Mesa AL in September 1987

Below:
Some Super King Airs are flown by small commuter airlines.

Above:
The Beech 1900 is used by commuter airlines, mainly in the USA. *Austin J. Brown*

British Aerospace HS748

Twin turboprop short range transport

Basic data for Hs748 Series 2B

Powerplant: Two Rolls-Royce Dart 552 of 2,280ehp
Span: 102ft 6in (31.23m)
Length: 67ft 0in (20.42m)
Max cruise: 278mph (447km/hr)
Passengers: 52 plus two crew

First aircraft flown: 24 June 1960
Production: Over 377 built, including 18 Series 1. Now replaced by ATP in production
Recent/current service with: 50 airlines worldwide including Air BVI, British Airways, Dan Air, DLT, Euroair, LIAT, Mount Cook, Philippine, Quebecair, Ryan Air, Zambia Airways and others

Recognition: Twin turboprops mounted above and forward of the low-set wing, with the undercarriage fairing beneath. Wings have dihedral and taper towards the tips. A broad, unswept fin and rudder with a dorsal fillet. Low set tailplane either side of the lower rear fuselage. Some versions have a large, rear freight loading door
Variants: The Series 1, 2 and 2A are all externally similar. The Series 2B has a 4ft (1.22m) extension to the wing span and an optional freight door

Below:
A twin Dart-engined HS748 Srs 1 operated by Dan Air. *Andrew March*

Above:
Air Atlantique was a new UK HS748 operator in 1988.

British Aerospace ATP

Twin turboprop regional airliner

Basic data for ATP

Powerplant: Two Pratt & Whitney Canada PW124 turboprops of 2,520shp
Span: 100ft 6in (30.63m)
Length: 85ft 4in (26.01m)
Max cruise: 306mph (492km/hr)
Passengers: 68 plus two crew

First aircraft flown: 6 August 1986
Production: 20-plus ordered — British Airways, British Midland, LIAT and Manx. First production delivery to BMA on 21 April 1988, entering airline service the following month
Recognition: Resembles a stretched HS748 with twin turboprops mounted above and forward of the low-set wings. The fuselage, circular in section, is stretched forward and aft of the wings. Redesigned swept fin and rudder and modified, more pointed nose

Below:
The BAe ATP has replaced the HS748, from which it was developed, in production.

British Aerospace Jetstream

Twin turboprop commuter airliner

Basic data for Jetstream 31

Powerplant: Two flat-rated Garrett TPE-331-10 turboprops of 940shp
Span: 52ft 0in (15.85m)
Length: 47ft 1½in (14.36m)
Max cruise: 303mph (488km/hr)
Passengers: 18 plus one/two crew

First aircraft flown: 28 March 1980; the original Handley Page Jetstream was flown on 18 August 1967
Production: Over 100 of the 200 ordered had been completed by 1988
Recent/current service with: Australian, Birmingham Executive, Contractair, Metro Airlines, Netherlines

Recognition: Twin turboprops mounted above and forward of the low-set wings. Circular fuselage with a long, pointed nose forward of the cockpit. Swept, tall fin and rudder with a triangular ventral extension. Circular cabin windows and a passenger door aft of the port wing
Variants: The original Handley Page Jetstream has Turbomeca Astazou XIV turboprops which are more slender and extend further forward of the wing, with a distinctive long spinner. A stretched Jetstream 41 is proposed

Below:
This Jetstream 31 is used by Birmingham Executive for European services.
Neil Hargreaves

CASA C-212 Aviocar

Twin turboprop commuter airliner and light transport

Basic data for C-212-200 Aviocar

Powerplant: Two Garrett TPE331-10 turboprops of 925shp
Span: 62ft 4in (19.0m)
Length: 44ft 9in (13.64m)
Max cruise: 220mph (354km/hr)
Passengers: 19 plus one/two crew

First aircraft flown: 26 March 1971
Production: Over 400 sold for civil and military use, including about a quarter under licence by Nurtanio in Indonesia
Recent/current service with: Numerous commuter airlines in the US and non-scheduled airlines in Europe
Recognition: Twin turboprops mounted forward of the high, straight wing. A rectangular fuselage with a short, pointed nose and upswept rear section. An angular fin and rudder with a large dorsal extension. Undercarriage fairings on the lower fuselage below the wing. Tailplane positioned below the fin and rudder on an extension of the fuselage
Variants: The initial production Series 100 had lower power turboprops; Series 200 is externally similar. The latest Series 300 is larger, with span increased to 67ft (20.4m) and the length to 53ft (16.15m), a modified nose and a reshaped rear fuselage in place of the rear loading ramp

Below:
The commuter CASA 212 has a box-like fuselage.

Above:
Latest version of the CASA 212 is the Series 300 with a modified wing and increased passenger accommodation.

CASA-Nurtanio CN-235

Twin turboprop regional airliner and light cargo transport

Basic data for CN-235

Powerplant: Two General Electric CT7-7A turboprops of 1,700shp
Span: 84ft 7½in (25.79m)
Length: 70ft 0½in (21.35m)
Max cruise: 277mph (446km/hr) at 18,000ft
Passengers: 44 plus two crew

First aircraft flown: 11 November 1983 (Spain), 31 December 1983 (Indonesia)
Production: Jointly manufactured by CASA and Nurtanio. Over 120 aircraft ordered by early 1988, half for military customers
For operation with: Aviaco, Deraya Air Transport, Merpatic and Pelita Air Services, but not in airline service by early 1988
Recognition: Twin turboprops set forward and under the high mounted, tapered wings with extended tips. Large undercarriage fairings set under the circular fuselage centre section; rear fuselage swept up with rear loading ramp. Tall, swept fin and rudder with a long dorsal fairing. Straight tailplane set below the fin and rudder

Below:
Developed jointly in Spain and Indonesia, the CN-235 is beginning to sell to regional airlines. *Andrew March*

Cessna 404 Titan

Twin piston engined light commuter transport

Basic data for Cessna 404 Titan

Powerplant: Two Continental GTS10-520-M of 375hp
Span: 46ft 4in (14.12m)
Length: 39ft 6¼in (12.05m)
Max cruise: 229mph (369km/hr)
Passengers: 10 including crew

First aircraft flown: 26 February 1975
Production: Over 360 built, when production ceased in 1985
Recent/current service with: Numerous small charter airlines worldwide
Recognition: Twin piston engines mounted above and forward of the straight, low-set, square-tipped wings. A rectangular fuselage with a long, round nose and large cabin windows extending from behind the cockpit to aft of the port-side passenger entry door. Swept fin and rudder with a small dorsal fillet. A straight tailplane with dihedral, low-set on the rear fuselage below the fin and rudder
Variants: There are three main variants of the Titan, designated according to function; Titan Ambassador — passenger version; Titan Courier — passenger/cargo version; Titan Freighter — light freight carrier. The latter two have large rear cargo doors. Although the Titan is the most extensively used Cessna twin by commuter and charter airlines, other types can be seen in airline livery:

Cessna 402B/C Businessliner — executive, and Utiliner — convertible. A low-wing, unpressurised, turbo-charged twin piston which is smaller than the Titan. It features wing-tip fuel tanks. Rectangular windows
Cessna 406 Caravan II — French-built with wings and landing gear of the Conquest II, fuselage of the 404 Titan and engine nacelles of the Conquest I
Cessna 414A Chancellor — pressurised fuselage of the 402's dimensions, with smaller oval cabin windows. A similar wing to the Titan
Cessna 421C Golden Eagle — pressurised fuselage with a new wing without tip-tanks. Earlier versions carry tip tanks. Oval windows
Cessna 425 Conquest I — turboprop variant of the pressurised Golden Eagle, originally called Corsair. Oval windows. Dihedral tail
Cessna 441 Conquest II — pressurised fuselage developed from 421C with turboprops. Rectangular ovalated windows. Dihedral tail

Above:
The Cessna 404 Titan is mainly operated by smaller air taxi, private charter and light freight companies.

Below:
The Cessna 421B is a pressurised twin with wing-tip fuel tanks as standard.
Andrew March

Convair CV-240, CV-340, CV-440 Metropolitan

Twin piston-engined short range airliner

Basic data for Convair CV-440

Powerplant: Two Pratt & Whitney R-2800-CB17 piston engines of 2,500hp
Span: 105ft 4in (32.11m)
Length: 81ft 6in (24.84m)
Max cruise: 300mph (483km/hr)
Passengers: 56 plus two/three crew

First aircraft flown: 8 July 1946 (CV-110); 16 March 1947 (CV-240)
Production: A total of 968 of all models (up to CV-440) for civil and military use
Recent/current service with: Over 60 remain in service in USA — General Aviation, Sundance Airlines, Trans Florida, Air Resorts Airlines, Renown Aviation
Recognition: Twin piston engines projecting forward of the tapered, low-set wing which has dihedral towards the tips. A circular fuselage with a shaped fin and rudder having a distinctive curved leading edge. The tailplane is set on the sides of the rear fuselage below the fin and rudder
Variants: CV-240 — original short-fuselage (74ft 8in [22.76m]) production version; CV-340 — has a 4ft 6in (1.37m) longer fuselage and broader chord wings; CV-440 — same fuselage as the 340 with modified engine cowlings and internal improvements. Nose lengthened by 2ft 4in (0.71m) when weather radar incorporated

Below:
The piston-engined Convair 440 serves with a handful of airlines, mainly as a light freighter. *Andrew March*

Above:
A well worn Convair 440 getting some attention on the ramp. *Neil Hargreaves*

Convair 580, 600 and 640

Twin turboprop short range airliner

Basic data for Convair 580

Powerplant: Two Allison 501-D31H turboprops of 3,750shp
Span: 105ft 4in (32.11m)
Length: 81ft 6in (24.84m)
Max cruise: 342mph (550km/hr)
Passengers: 56 plus two/three crew

First aircraft flown: 19 January 1960 (Convair 580)
Production: Total of 243 converted to turboprop standard from earlier models
Recent/current service with: Several US provincial airlines including Air Ontario, Aspen, Northwest and Quebecair. Partnair and European Air Transport are the only European operators with 580s

Recognition: As Convair CV-240/440 with twin turboprops replacing the earlier piston engines. This gives a slimmer engine profile. The 580, 600 and 640 series all have the longer nose incorporating a weather radar aerial
Variants: Convair 580 has Allison 501-D13H turboprops; Convair 600 is a RR Dart engined conversion of the CV-240, and the Convair 640 is a similarly engined conversion of the CV-340/440

Below:
Allison turboprops give a slimmer engine profile to this Continental Express Convair 580.

Above:
The re-engined Convair 580 is widely used by commuter airlines in North America. *Neil Hargreaves*

De Havilland Canada Dash 8

Twin turboprop regional airliner

Basic data for Dash 8 Series 100

Powerplant: Two Pratt & Whitney Canada PW120 turboprops of 2,000shp
Span: 85ft 0in (25.91m)
Length: 73ft 0in (22.25m)
Max cruise: 311mph (500km/hr)
Passengers: 36 with two crew

First aircraft flown: 20 June 1983; entered service October 1984
Production: By early 1988 over 70 aircraft delivered from 145 ordered
Current service with: 16 airlines including Tyrolean Airways and DLT in Europe and many US commuter airlines
Recognition: Narrow profile turboprops set underneath high-set, narrow chord, unswept wings. Circular fuselage section which sweeps up to a broad, slightly swept, rectangular fin and rudder. Dorsal extension reaches forward to the trailing edge of the wing. Straight tailplane set on top of the fin. Streamlined nose with a continuous line down from the cockpit
Variants: Series 300 has a longer (11ft 6in) fuselage to accommodate up to 56 passengers, and Pratt & Whitney PW123 turboprops. The Dash 8M is a military version of the Series 100

Below:
A colourful Horizon Air Dash 8 in service in North America. *Neil Hargreaves*

Above:
Twin-engined Dash 8s have STOL performance.

De Havilland Canada Twin Otter

Twin turboprop commuter transport

Basic data for Twin Otter Series 300

Powerplant: Two Pratt & Whitney PT6A-27 turboprops of 620shp
Span: 65ft 0in (19.81m)
Length: 51ft 9in (15.77m)
Max cruise: 209mph (336km/hr)
Passengers: 20 plus two crew

First aircraft flown: 20 May 1965
Production: More than 830 produced by 1988, including military users
Recent/current service with: Brymon and Loganair in the UK and many feeder and commuter airlines worldwide
Recognition: Twin turboprops situated below and forward of the high-set 'plank' wing. A fixed tricycle undercarriage with the main wheels attached to the lower fuselage below the wing. Bracing struts extend from the undercarriage fairings to the engines. A flat-sided cabin with small, square windows set high up. Tapered rear fuselage and a long nose forward of the cockpit. The fin and rudder is tall, slightly tapered and square cut at the top, with the tailplane mounted near to the base
Variants: Series 100 had a short nose. Series 200 features an extended nose, and the Series 300 has the longer nose and uprated engines

Below:
The last example of the rugged STOL Twin Otter was produced in 1988.

Dornier Do228

Twin turboprop commuter airliner

Basic data for Dornier Do228 Series 200

Powerplant: Two Garrett TPE-331-5 turboprops flat-rated at 715shp
Span: 55ft 7in (16.94m)
Length: 54ft 3in (16.54m)
Max cruise: 230mph (370km/hr)
Passengers: 19 plus two crew

First aircraft flown: 28 March 1981 (Series 100) and 9 May 1981 (Series 200)
Production: Over 100 aircraft delivered by early 1988. Also being built under licence in India
Recent/current service with: 42 airlines worldwide including Air Vendee, AS Norving, Druk Air, FRL, Olympic, Somali and Suckling
Recognition: Small turboprops mounted on the leading edge of the 'new technology' wing which has a straight trailing edge and curved leading edge with pointed wing-tips. The square, slab-sided fuselage is positioned below the wing. Forward of the cockpit the flat-bottomed nose is shaped downwards to give a 'drooped' appearance. The angular fin and rudder has a large dorsal fillet extending to the rear of the wing fairing. The flat tailplane is mounted below the rudder, extending beyond the fuselage cone. The main undercarriage retracts into lower fuselage fairings
Variants: Series 100 has a fuselage of 49ft 3in (15.01m) for 15 passengers; the Series 200 has a 5ft (1.52m) longer fuselage for 19 passengers

Below:
The Do228-200 has a square section fuselage and a crescent-shaped wing profile. *Neil Hargreaves*

Douglas DC-3 Dakota

Twin piston-engined short range transport

Basic data for Douglas DC-3

Powerplant: Two Pratt & Whitney R-1830-92 Twin Wasps of 1,200hp
Span: 95ft 0in (28.96m)
Length: 64ft 6in (19.66m)
Max cruise: 194mph (312km/hr)
Passengers: 32 plus two crew

First aircraft flown: 17 December 1935
Production: 10,962 built of which only 458 were originally for civilian use. Just over 300 remain in commercial use today
Recent/current service with: Airlines worldwide including over 110 operators in the USA, over 20 in the Middle East, Far East and Africa and four in Europe including Air Atlantique in the UK
Recognition: The engines are positioned forward of the wing, close in to the fuselage. A low-set wing with swept leading edges and dihedral outboard of the engines, straight trailing edge and almost pointed wing-tips. The oval fuselage is set above the wing with a distinctive rounded nose. The broad fin and rudder has a curved top. Mounted on either side of the fuselage below the fin, the tailplane has a swept front edge and straight trailing edge. The main undercarriage retracts into the lower part of the engine cowlings leaving one-third of the wheel exposed; the tailwheel does not retract
Variants: Various modifications have been made to the DC-3 for VIP use with modified windows, powerplant refinements, nose radar and fully enclosed housings for the main wheels; cargo versions feature large 'double' doors on the port side of the rear fuselage

Below:
The DC-3 can still be found throughout the world as a short range passenger and cargo work-horse. *Andrew March*

Above:
This Stellair DC-3 is one of three in service during 1988 with the French regional passenger airline. *Colin Wood*

Embraer EMB-110 Bandeirante

Twin turboprop commuter airliner

Basic data for EMB-110P2 Bandeirante

Powerplant: Two Pratt & Whitney PT6A-34 turboprops of 750shp
Span: 50ft 3in (15.32m)
Length: 49ft 6½in (15.01m)
Max cruise: 244mph (393km/hr)
Passengers: 56 plus two crew

First aircraft flown: 26 October 1968 (military version); 16 April 1973 (EMB-110C) and 3 May 1977 (EMB-110P2)
Production: Over 475 delivered worldwide to operators in 35 countries
Recent/current service with: Jersey European in the UK, Brit Air in France and many commuter airlines in the USA
Recognition: Twin turboprops project well forward of the low-set, straight wings with square wing-tips. Straight fuselage sides with large rectangular cabin windows. The passenger doors are behind the cockpit and to the rear of the wing. A slightly swept angular fin and rudder has a dorsal extension; there is also a small ventral fin below the rear fuselage. The straight tailplane is set below the rudder on the fuselage cone
Variants: The shorter EMB-110C is in limited use. The main variants are the stretched EMB-110P1 with a large rear cargo door and the EMB-110P2 with fore and aft passenger doors

Below:
This Jersey European EMB-110P2 Bandeirante has the fore and aft passenger doors of later models.

Embraer EMB-120 Brasilia

Twin turboprop regional airliner

Basic data for EMB-120

Powerplant: Two Pratt & Whitney Canada PW118 turboprops of 1,800shp
Span: 64ft 10¾in (19.78m)
Length: 65ft 7¼in (20.00m)
Max cruise: 345mph (555km/hr) at 20,000ft
Passengers: 30 with two crew

First aircraft flown: 27 July 1983
Production: First delivery to US commuter airline in July 1985 and DLT in Germany in January 1986. 62 of over 128 ordered in service by early 1988
Current service with: 10 airlines in Europe and US in 1987, including Air Littoral, Norsk Air, Air Midwest, Atlantic Southeast, Norsk Air, Skywest AL and West Air
Recognition: Two turboprops projecting forward from the low-set, straight wings. The wings have a slight dihedral and the leading edge is swept inboard of the engines. Slim, circular fuselage with a large swept fin and rudder with dorsal projection. The swept tailplane is located on top of the fin. Pointed nose with sharply raked large cockpit windscreens

Below:
The sleek-looking Embraer Brasilia has the highest cruising speed in this class of regional airliners. *Embraer*

Fairchild Metro/Expediter

Twin turboprop commuter airliner

Basic data for Metro III

Powerplant: Two Garrett TPE-331-11U-601G turboprops of 1,000shp
Span: 57ft 0in (17.37m)
Length: 59ft 4¼in (18.09m)
Max cruise: 317mph (511km/hr)
Passengers: 19 plus two crew

First aircraft flown: 26 August 1969
Production: Over 380 of all variants delivered by 1988 including 20 Expediters
Recent/current service with: Many US commuter airlines but only a limited number in Europe including Air Vendée, CAL, Crossair and Metro Airways
Recognition: A long, slender, circular section fuselage is mounted above the slightly tapered wings. The twin turboprops extend well forward of the wings and are close into the fuselage. The raked fin and rudder appears small on the longer fuselage. The tailplane, set above the fuselage on the dorsal extension, is swept sharply
Variants: Originally developed by Swearingen from the Beech Queen Air as the Merlin II, the pressurised Metro was built by Fairchild. The Metro II had larger windows and other refinements. The Metro III features a new wing of 10ft 9in (3.28m) greater span; the Expediter is an all-cargo version with a strengthened cabin floor

Below:
The Fairchild Metroliner has a circular section fuselage and slightly tapering wings.

Fokker F27 Friendship/Fairchild FH227

Twin turboprop short range transport

Basic data for F27 Mk 500 Friendship

Powerplant: Two Rolls-Royce Dart 536-7R turboprops of 2,320shp
Span: 95ft 2in (29.01m)
Length: 77ft 3½in (23.56m)
Max cruise: 298mph (480km/hr)
Passengers: 60 plus two crew

First aircraft flown: 24 November 1955
Production: Over 786 built by 1987 including 206 in USA by Fairchild. Fokker production includes 85 Mk 100, 138 Mk 200, 13 Mk 300, 218 Mk 400/600, 112 Mk 500. Fairchild built 128 F27s and 78 FH227s
Recent/current service with: Air UK, British Midland Airways and 18 other European airlines and many provincial operators worldwide. FH227 operated by Delta AT and TAT in Europe and ALFA, Britt Airways, Canadian AL and TABA in the Americas

Recognition: Twin turboprops set below the high, straight wing. A slender oval section fuselage with a pointed nose and tail; the cabin windows are distinctively oval shaped. A tall fin and rudder with a large dorsal extension. The small tailplane is set either side of the base of the rudder
Variants: F27 Mk 200 (F27A) has improved powerplants; F27 Mk 300 (F27B) 'Combiplane' has a large forward cargo door; F27 Mk 400 (F27M) has improved powerplants; F27 Mk 500 has 5ft (1.52m) longer fuselage and a large forward cargo door; F27 Mk 600 retains the shorter fuselage but has the other improvements of the Mk 500; FH227B to E has a 6ft (1.83m) longer fuselage of 83ft 8in (25.5m) and other improvements

Below:
Air UK has a large fleet of Fokker F27s including this Series 200.

Fokker 50

Twin turboprop short range transport

Basic data for Fokker 50

Powerplant: Two Pratt & Whitney Canada PW124 turboprops of 2,240shp
Span: 95ft 1¾in (29.00m)
Length: 82ft 7¾in (25.25m)
Max cruise: 332mph (535km/hr) at 20,000ft
Passengers: 50 plus two crew

First aircraft flown: 28 December 1985
Production: First production aircraft flown February 1987, with orders for 44 by early 1988
Current orders with: Ansett, DLT, Busy Bee, Austrian AL, Maersk Air, Malaysia and Sudan AW. Entered service with DLT in late 1987
Recognition: Little different from the F27-500 from which it is derived. Twin turboprops set below the high, straight wings. A slender, oval section fuselage with a pointed nose and tail. A tall fin and rudder with a large dorsal extension. The small tailplane is set either side of the base of the rudder. Fokker 50 has smaller rectangular cabin windows than F27-500

Below:
The F27 has been replaced in production by the improved Fokker 50 which features new powerplants with six-bladed propellers. *Neil Hargreaves*

GAF Nomad

Twin turboprop commuter/light transport

Basic data for Nomad N24

Powerplant: Two Allison 250-B17B turboprops of 200shp
Span: 54ft 0in (16.46m)
Length: 47ft 1in (14.35m)
Max cruise: 193mph (311km/hr)
Passengers: 16 plus two crew

First aircraft flown: 23 July 1971 (N22) and 17 December 1975 (N24)
Production: 170 when deliveries stopped in 1984
Recent/current service with: Small airlines in the US, Caribbean and Pacific areas
Recognition: Twin turboprops set below and forward of the high 'plank' wing. A box fuselage with slightly bulged, large cockpit windows; rectangular cabin window and large doors on the port side to the rear of the wing. An upswept rear fuselage with a very tall fin and rudder. Undercarriage sponsons on stubs from the lower edge of the fuselage with bracing struts to the wings outboard of the engines. The rectangular tailplane is set near to the base of the fin
Variants: The N22 is a 12-seater with a short fuselage (41ft 2½in/ 12.56m) while the 16-seater N24 is nearly 6ft (1.83m) longer (47ft 1in/ 14.35m)

Below:
The Australian designed GAF Nomad is identified by its angular fuselage and deep flight deck windows.

Handley Page Herald

Twin turboprop short range transport

Basic data for Herald Series 200

Powerplant: Two Rolls-Royce Dart 527 turboprops of 2,150shp
Span: 94ft 9in (28.88m)
Length: 75ft 6in (23.01m)
Max cruise: 274mph (441km/hr)
Passengers: 56 plus two crew

First aircraft flown: 25 August 1955 (piston engined); 11 March 1958 (Dart engined)
Production: 50 built (two prototypes, four Series 100, 36 Series 200 and eight Series 400) up to 1968 when production was concluded
Recent/current service with: Aerosucke, Aerovias Guatemale, British Air Ferries, Channel Express, Euroair and Skyguard

Recognition: Engines set below and forward of the high, unswept wing which has noticeable dihedral. The circular fuselage has a high-set cockpit. The broad, angular fin and rudder has a dorsal extension stretching forward towards the trailing edge of the wing. Large, double, port-side cabin doors are located at the rear of the fuselage, forward of the tailplane
Variants: Series 100 — shorter fuselage (71ft 11in/21.92m); Series 200 — 3ft 7in (1.09m) extension to the fuselage (75ft 6in/23.01m); Series 400 — a military version of the Series 200

Below:
Channel Express operates a fleet of Heralds on cargo flights between the UK and the Channel Islands.

LET L-410

Twin turboprop light transport

Basic data for LET L-410UVP-E

Powerplant: Two Walter M601B turboprops of 750shp
Span: 65ft 6in (19.98m)
Length: 47ft 5½in (14.47m)
Max cruise: 227mph (365km/hr)
Passengers: 15 plus two crew

First aircraft flown: 16 April 1969; L-410UVP first flown 1 November 1977
Production: Over 720 delivered by early 1987
Recent/current service with: Aeroflot, CSA and Slov-Air
Recognition: Engines mounted below and forward of the high-set wing. A short, oval section fuselage with a long tapered nose forward of the cockpit. Seven rectangular cabin windows, with a large cabin door just aft of the wing. Large undercarriage fairings extend out from the lower fuselage. A slightly swept, tall, angular fin and rudder with a ventral extension below the tail cone. The straight, dihedralled tailplane is mounted on the fin above the fuselage
Variants: Early production L-410A had Pratt & Whitney PT6A turboprops, the L-410M introduced Walter M601 powerplants. The L-410UVP with a slightly larger fuselage, extended wing-tips, taller fin and rudder and other detailed changes was produced until 1985. Now in production is the L-410UVP-E which accommodates four more passengers, has new engines driving five-blade propellers and wing-tip fuel tanks

Below:
The LET L-410UVP is widely used in Eastern Bloc countries as a utility transport.

Pilatus Britten-Norman Islander

Twin piston/turboprop light transport

Basic data for BN-2T Turbine Islander

Powerplant: Two Allison 250-B17C turboprops of 320shp
Span: 49ft 0in (14.94m) (53ft [16.15m] with extended wing-tips)
Length: 35ft 7¾in (10.86m)
Max cruise: 212mph (341km/hr)
Passengers: Nine plus one crew

First aircraft flown: 13 June 1965 (BN-2 Continental IO-360); 6 April 1977 (Turbo Islander Lycoming LTP101); 2 August 1980 (BN-2T Allison 250)
Production: Over 1,080 delivered by 1988. Production lines are now located in Romania AIAV Bucharest and the Philippines PADC only
Recent/current service with: Many commuter and air taxi airlines worldwide
Recognition: Engines mounted below and forward of the straight, high-set 'plank' wing. Fixed tricycle undercarriage with the main wheels on an extended, faired leg at the rear of the engines and the nosewheel situated well forward below the nose cone. Slab-sided rectangular fuselage with a level top surface and gently raked lower surface aft of the wing, two port-side cabin entry doors and large rectangular cabin windows. The tall, angular fin and rudder has a small dorsal fillet; the straight tailplane is mounted on top of the fuselage, below the rudder. The extended wing-tips on some aircraft have a distinctive conical camber
Variants: The BN-2, BN-2A and BN-2B are all broadly similar piston-engined versions. The latter two variants could have optional extras which included a lengthened nose forward of the cockpit and/or 4ft (1.22m) extended wing-tips. The BN-2T has smaller Allison 250 turboprops in place of the Lycoming series piston engines

Below:
The rugged Islander continues to be sold to air taxi and commuter operators worldwide.

Piper Navajo & Chieftain

Twin piston/turboprop light transport

Basic data for Chieftain

Powerplant: Two Lycoming LT10-540 of 350bhp
Span: 40ft 8in (12.40m)
Length: 36ft 8in (11.18m)
Max cruise: 254mph (409km/hr) at 20,000ft
Passengers: 10 plus two crew

First aircraft flown: 30 September 1964 (PA31 Navajo)
Production: Some 2,500 of the Navajo series built
Recent/current service with: Air taxi and commuter airlines world-wide and particularly in the USA
Recognition: Engines mounted above the straight wings, projecting forward from the leading edge and slightly to the rear of the wings which are low-set. A distinctive leading edge fillet between the engines and the fuselage. Short, oval section fuselage with a long conical-shaped nose and large rectangular cabin windows extending from the cockpit to the passenger door aft of the wing. Tall, swept fin and rudder with the tailplane mounted either side of the rear fuselage cone
Variants: The original eight-seat PA-31 Navajo had 300hp engines and a short fuselage. The Chieftain which was 2ft (0.61m) longer, had a modified nose and cabin, improved passenger access and 350hp engines

Below:
More powerful than early Piper twins, the Navajo Chieftain can carry extra passengers for short commuter and air taxi services.

Saab 340

Twin turboprop regional airliner

Basic data for Saab SF340

Powerplant: Two General Electric CT7-5A turboprops of 1,735eshp
Span: 70ft 4in (21.44m)
Length: 64ft 9in (19.72m)
Max cruise: 300mph (483km/hr)
Passengers: 34 plus two crew

First aircraft flown: 25 January 1983
Production: 132 ordered by early 1988 with 95 in service
Current service with: 20 airlines mainly in Europe and USA, including Crossair, Europe AirService, Manx, Norving and Swedair also Air Midwest, Chicago Air, Comair, Metro, and Republic Express
Recognition: Slim engines projecting forward of the low, straight wing which is set mid-way along the circular-section fuselage. A swept fin and rudder with a dorsal fillet projecting forward to the cabin windows. The small, dihedralled tailplane is mounted either side of the tail cone below the fin. The cockpit windscreen slopes down to the nose cone in a continuous line

Below:
Crossair was the lead customer for the Saab 340. *Andrew March*

Above:
A 34-seat regional airliner, the Saab 340 meets the latest safety requirements. *Andrew March*

Shorts Skyvan

Twin turboprop short range transport

Basic data for Skyvan Series 3

Powerplant: Two Garrett TPE 331-201 turboprops of 715shp
Span: 64ft 11in (19.79m)
Length: 40ft 1in (12.21m)
Max cruise: 203mph (327km/hr)
Payload: 4,600lb or 18 passengers and two crew

First aircraft flown: 17 January 1963 (Series 1) with Continental piston engines; 2 October 1963 (Series 2) with Astazou turboprops; and 15 December 1967 (Series 3) with Garrett turboprops
Production: Over 150 in service, including a handful of Series 2s
Recent/current service with: Regional airlines in Asia and Pacific areas in particular
Recognition: A distinctive angular appearance with a short, square-sided fuselage and sharply up-swept rear section. The narrow chord, straight wing is set on top of the fuselage. Small engines are close into the fuselage and project forward and below the wing. A bracing strut extends from the mid-point of the wing to the fixed undercarriage which projects from the lower fuselage. Twin, rectangular fins and rudders are located at the ends of the tailplane which is situated at the extremity of the fuselage. The sharply-angled nose incorporates the front cockpit windscreen. The fixed nosewheel is located below the nose cone
Variants: Series 1 — piston engines; Series 2 — long, slender Astazou turboprops; Series 3 — stubby Garrett turboprops

Below:
The Skyvan has a fixed undercarriage and a short angular fuselage.

Shorts 330

Twin turboprop regional airliner

Basic data for Shorts 330-200

Powerplant: Two Pratt & Whitney PT6A-45R of 1,198shp
Span: 74ft 8in (22.76m)
Length: 58ft 0½in (17.69m)
Max cruise: 219mph (352km/hr)
Passengers: 30 plus two crew

First aircraft flown: 22 August 1974 (SD3-30)
Production: 120 delivered mid-1987
Recent/current service with: British Air Ferries, Guernsey Airlines and Jersey European in the UK, 15 commuter airlines in N/S America and a handful in the Far East as well as a number of North American airlines
Recognition: Larger development of the Skyvan with the same general features including a slab-sided fuselage and twin fins and rudders. The top of the fuselage is noticeably humped and the nose and the rear sections are more streamlined. The undercarriage retracts, with the main wheels entering a fairing located either side of the fuselage below the wings
Variants: Prototype SD3-30, renamed the 330-100 has 1,173shp PT6A-45B turboprops while the 330-200 has uprated 1,198shp PT6A-45Rs. The Sherpa 330-UTT is a military utility development of the 330-200 in service with the USAF as the C-23A

Below:
Developed from the Skyvan, the Shorts 330 has a distinctive straight, narrow chord wing and twin fins.

Shorts 360

Twin turboprop regional airliner

Basic data for Shorts 360

Powerplant: Two Pratt & Whitney PT6A-45R of 1,198shp
Span: 74ft 8in (22.76m)
Length: 70ft 6in (21.49m)
Max cruise: 242mph (389km/hr)
Passengers: 36 plus two crew

First aircraft flown: 1 June 1981
Production: Order book standing at 125 by 1988 with 117 in service with 30 airlines
Recent/current service with: Air UK, British Midland, Jersey European, Loganair, Manx Airlines and Aer Lingus together with 12 US commuter airlines
Recognition: A refined development of the Shorts 330 with a 3ft (0.91m) longer fuselage forward of the wing, and a distinctive, tall, single fin and rudder, set on top of the tapered upswept rear fuselage. The wing-shape, forward fuselage, bracing struts, retractable undercarriage and powerplants are similar to the Shorts 330
Variants: Externally all versions are similar. The latest 360-300 has P&W Canada PT6A-67R turboprops driving six-blade propellers and internal improvements

Below:
The latest version of the Shorts 360 is the 300 series, here in the colours of Capital Airlines.

Airliners in brief

Aero Spacelines/UTA Super Guppy

Four turboprop outsize cargo airliner

Powerplant: Four Allison 501-D22C turboprops of 4,508eshp
Span: 156ft 3in (47.62m)
Length: 143ft 10in (43.84m)

Based on the structure of the Boeing Stratocruiser/KC-97, two conversions have been completed by Aero Spacelines and two by UTA in France. All serve on the Airbus major component delivery service, 'Skylink', to Toulouse from Airbus manufacturing partners throughout Europe

Super Guppy F-BTGV. *Andrew March*

Airbus A330

Twin turbofan-engined long range airliner

Powerplant: Two General Electric CF6-80C2 turbofans of 64,000lb st
Span: 192ft 5in (58.6m)
Length: 208ft 10in (62.6m)

Launched in June 1987, this 375-seater is scheduled to enter service in late 1993

Airbus A330. *Artist's impression*

Airbus A340

Four turbofan-engined very long range airliner

Powerplant: Four General Electric SNECMA CFM-56-5C turbofans, of 31,200lb st
Span: 192ft 5in (58.6m)
Length: 194ft 10in (59.4m) Series 200, 208ft 10in (63.6m) Series 300

Launched concurrently with the A330, this four-engined airliner will be marketed to highlight its very long range capability. Airline service is programmed for late 1992 after a 12 month flight certification programme

Airbus A340. *Artist's impression*

Cessna 208 Caravan 1 N9738B.

Cessna 208 Caravan I

Single turboprop light freight transport

Basic data for Cessna 208B

Powerplant: One Pratt & Whitney PT6A-114 turboprop
Span: 52ft 1¼in (15.88m)
Length: 41ft 7in (12.67m)
Max cruise: 212mph (341km/hr)
Passengers: Nine in passenger configuration

First aircraft flown: 9 December 1982 (208), 3 March 1986 (208B)
Production: Over 180 built in total, including 125 for Federal Express (200 ordered)
Recent/current service with: Federal Express for night parcel service and operators worldwide

Recognition: Single turboprop engine on long fuselage with no side windows (Fed Ex version) and swept fin with deep dorsal fin. Long baggage container below fuselage (optional). Fixed tricycle undercarriage. High wing with high-aspect ratio strut braced to fuselage
Variants: 208A version with six side windows or side cockpit window only on freighter version. 208B with 4ft (1.22m) fuselage extension, passenger or freighter configuration. Floatplane version is also available

Convair Coronado

Four turbofan-engined medium/long range transport

Powerplant: Four General Electric CJ805-23B turbofans of 16,050lb st
Span: 120ft 0in (36.58m)
Length: 139ft 2½in (42.43m)

Following the demise of Spantax in 1988, the remaining Coronados are stored at Palma, Majorca

Curtiss C-46 Commando

Twin piston-engined short range transport

Powerplant: Two Pratt & Whitney R-2800-51M1 Double Wasp piston engines of 2,000bhp
Span: 108ft 0in (32.92m)
Length: 76ft 4in (23.27m)

Of the 2,882 C-46s built 37 remained in airline service with South American airlines early in 1988 as cargo transports

Curtiss C-46 Commando N611Z. *Paul Gingell*

Fairchild C-123K Provider

Twin piston-engined transport

Powerplant: Two Pratt & Whitney R-2800-99W piston engines of 2,500hp and two Fairchild J44-3 turbojets of 1,000lb st
Span: 110ft (33.53m)
Length: 76ft 3in (24.92m)
Max cruise: 228mph (367km/hr) with jets; 173mph (278km/hr) normal

First flight: 14 October 1949 (XC-123); 27 May 1966 (C-123K)

A few ex-military transports remain in service in the USA but may be augmented when the Royal Thai Air Force retires its aircraft

C-123K Provider N689SM.

Grumman Gulfstream I

Twin turboprop-engined commuter airliner

Powerplant: Two Rolls-Royce Dart 529-8X turboprops of 2,210eshp
Span: 78ft 4in (23.88m)
Length: 63ft 9in (19.43m)

Mainly used as a corporate aircraft but becoming used more regularly as a commuter airliner

Gulfstream 1 G-BRWN. *Andrew March*

Grumman G-73T Turbo Mallard

Twin turboprop amphibious passenger transport

Powerplant: Two Pratt & Whitney PT6A-34 turboprops of 750shp
Span: 66ft 8in (20.32m)
Length: 48ft 4in (14.73m)
Max cruise: 190mph (306km/hr)
First flight: 1946 (G-73)

In limited use with Chalks Airline from Miami to the Bahamas and with Virgin Airlines Seaplanes Shuttle. The earlier piston-engined Mallard also serves the latter airline and other small companies

Turbo Mallard N2974.

NAMC YS-11

Twin turboprop regional airliner

Powerplant: Two Rolls-Royce Dart 542-10K turboprops of 3,060shp
Span: 105ft (32.00m)
Length: 86ft 3½in (26.30m)

About 110 of the 182 YS-11s built by the Japanese consortium remain in airline service

NAMC YS-11 N187P. *Paul Gingell*

Piper Aztec

Twin piston-engined light business and commuter aircraft

Powerplant: Two Lycoming T10-540-C1A piston engines of 250hp
Span: 37ft 2½in (11.34m)
Length: 31ft 2¾in (9.52m)

Widely used by small charter and commuter airlines throughout Europe and the Americas

Piper Aztecs are flown by air taxi and other small airlines.

Vickers Vanguard/Merchantman

Four turboprop short/medium range transport

Powerplant: Four Rolls-Royce Tyne 512 of 5,545eshp
Span: 118ft 7in (36.14m)
Length: 122ft 10½in (37.45m)

44 aircraft built; only nine remain in service, currently with Air Bridge Carriers, Elan Air and Intercargo

The Merchantman remains in limited service. *Alan J. Wright*

Similar shapes

Seen from the same angle many airliners can appear very much alike and it becomes quite difficult to tell one type from another. Here is a selection of photographs of airliners which are superficially similar together with a key to their individual recognition features to help you distinguish them.

Lockheed TriStar 100

McDonnell Douglas DC-10-30

Key recognition feature: The third, fin mounted rear engine. The DC-10's engine is positioned clear of the rear fuselage and has a straight through jet pipe with a small extension of the fin on the top surface. The TriStar's third engine intake is moulded on to the top of the fuselage and the front of the fin. It exhausts through the fuselage tail cone aft of the tailplane

Other features: The TriStar has a more shaped nose than the DC-10 and a broader chord fin

Boeing 707-323C

McDonnell Douglas DC-8-63

Key recognition feature: The DC-8 has a slimmer looking fuselage with a longer nose forward of the cockpit

Other features: The Boeing 707 has a pitot tube projecting forward from the top of the fin; the DC-8's fin is more swept and tapered than the 707's. The DC-8 has fewer, larger windows which are more widely spaced than the 707's

Boeing 767-258

Airbus A300 B4

Key recognition feature: The number of spoilers extending below the wing. The Airbus A300 has five very pronounced underwing fences/flap fairings while the Boeing 767 has only three

Other features: The Airbus has a more circular section fuselage than the 767 and a straight top to the rear fuselage. the 767's fin is taller and narrower chord than the A300's and it has a more triangular shaped tailplane

Airbus A300 B2

Airbus A310-203

Key recognition feature: The shorter fuselage of the A310 which gives it a more 'dumpy' appearance. The A300's fuselage is noticeably longer forward of the wing

Other features: The A310 has three underwing fences/flap fairings outboard of each engine while the A300 has four

Boeing 757-28A

Boeing 767-204

Key recognition feature: The longer looking 'pencil' fuselage of the 757 compared with the 'fatter' fuselage of the 767. The 757 has a much longer fuselage section forward of the wing than the 767

Other features: The nose of the 757 has a flatter underside which gives it a more 'dropped' appearance than the 767's more rounded look

Tupolev Tu-154B-1

Boeing 727-228

Key recognition feature: The Tu-154 has two large fairings extending to the rear of the wing to accommodate the undercarriage when retracted. The Boeing 727 retracts its main wheels into fuselage fairings

Other features: The 727's fin is more swept than the Tu-154's and does not have a bullet fairing projecting forward from the top. The 727's third engine's jet-pipe extends further aft of the fuselage cone than the 154's which looks cut away. The Tu-154 has a more slender, pointed nose than the 727

Tupolev Tu-134A

McDonnell Douglas DC-9-82

Key recognition feature: The Tu-134A has a very different wing than the DC-9; it is more sharply swept with a broader chord. From the front it appears to have anhedral. It does not have any wing leading edge slats and has two large fairings extending from the trailing edge to house the main undercarriage

Other features: The T-tails are quite different. The Tu-134's fin has a broader chord, the tailplane is more sharply swept and large bullet fairing extends fore and aft. The DC-9 has a blunter nose when compared with the Tu-134's more pointed profile

Fokker F28 Fellowship 4000

McDonnell Douglas DC-9-15

Key recognition feature: The F28's wing is much less swept and broader in chord than the DC-9's and it has two pronounced fences/ flap fairings on each trailing edge

Other features: The F28 has a dorsal extension to the fin and a rectangular end to the fuselage which opens as air brakes. The DC-9 has neither of these, its tail cone being pointed. The F28 also has a curved leading edge to the top of the fin which extends above the tailplane

Handley Page Herald 201 *Alan Wright*

Fokker F27 Friendship

Key recognition feature: The nose section clearly distinguishes the two turboprop airliners. The FH227/F27 Friendship has a longer, more slender nose; the Herald's nose forward of the cockpit slopes down much more steeply than the FH227/27's, to give a blunter appearance

Other features: The Herald's tailplane is lower set and it has a broader chord fin and rudder than the FH227/F27

Shorts 330

Shorts 360

Key recognition feature: The Shorts 330 has twin, rectangular shaped fins and rudders while the Shorts 360 has a tall, single fin and rudder

Other features: The Shorts 360 has a longer fuselage forward of the wings, additional cabin windows and a longer, more gently tapered rear fuselage than the 330

Dornier 228 Series 100 *Andrew March*

Dornier 228 Series 200 *Andrew March*

Key recognition feature: The Series 200 has a noticeably longer fuselage, fore and aft of the wing
Other features: The longer Series 200 has two additional cabin windows to the 100's six on each side of the fuselage

Index